X-PLANES 12

DOUGLAS D-558

D-558-1 Skystreak and D-558-2 Skyrocket

Peter E. Davies

SERIES EDITOR TONY HOLMES

OSPREY
PUBLISHING

OSPREY PUBLISHING
Bloomsbury Publishing Plc
PO Box 883, Oxford, OX1 9PL, UK
1385 Broadway, 5th Floor, New York, NY 10018, USA
E-mail: info@ospreypublishing.com
www.ospreypublishing.com

OSPREY is a trademark of Osprey Publishing Ltd

First published in Great Britain in 2019

A catalog record for this book is available from the British Library.

ISBN: PB 9781472836212; eBook 9781472836229;
ePDF 9781472836205; XML 9781472836236

19 20 21 22 23   10 9 8 7 6 5 4 3 2 1

Artwork by Adam Tooby
Index by Zoe Ross
Originated by PDQ Digital Media Solutions, UK
Printed and bound in India by Replika Press Private Ltd.

Osprey Publishing supports the Woodland Trust, the UK's leading woodland
conservation charity.

To find out more about our authors and books visit www.ospreypublishing.com.
Here you will find extracts, author interviews, details of forthcoming events and
the option to sign up for our newsletter.

Acknowledgments
Thanks are due to Fred Johnsen and
Terry Panopalis for providing
photographic images.

Front Cover
On August 31, 1953 Lt Col Marion Carl
flew D-558-2 BuNo 37974 in an
attempt to add a speed record to the
new world altitude record of 83,235ft
that he had set on August 21. After an
air-launch in this all-rocket version of
the D-558-2 he reached Mach 1.5, but
the aircraft became laterally unstable.
In a final attempt on September 2, he
was able to push the speed to Mach
1.728. Carl had previously set a world
speed record of 650.6mph in D-558-1
BuNo 37970 on August 25, 1947.
(Cover artwork by Adam Tooby)

X PLANES

CONTENTS

CHAPTER ONE

INTRODUCTION

As the performance of fighter aircraft steadily advanced during World War II, pilots began to face increasing difficulty in controlling them at high speeds. A diving chase after an enemy machine by a pilot flying a fighter such as a P-38, P-51, or late-series Spitfire could often develop into a terrifying situation where the flying controls would not work, the aircraft began to vibrate severely, and (with luck) it might finally pull out of the dive, but at a dangerously low altitude. It was an experience that NACA (National Advisory Committee for Aeronautics) test pilot John Griffith described as trying to operate a control column that "felt like it was cast in about two feet of concrete."

Pilots were encountering the phenomenon known as compressibility, which had been studied theoretically for many years following wind tunnel work on propeller airfoils by Frank Caldwell and Elisha Fales at the Army Air Service Engineering Division in 1918. Their findings were confirmed by Dr Lyman Briggs and Dr Hugh Dryden in 1926, but only experienced in flight when aircraft began to exceed 400mph. The problem had been taken as evidence of the impossibility of traveling faster than the speed of sound because the symptoms of compressibility and loss of control began to appear when the airflow, passing over wings at air speeds exceeding 450mph, actually reached supersonic speed locally over some areas of the wings.

When it became clear in 1942 that Germany was developing high-speed jet aircraft, Gen Henry H. "Hap" Arnold, chief of the USAAF (United States Army Air Force), engaged a Scientific Advisory Board led by the respected and persuasive Hungarian high-speed aerodynamics expert Dr Theodore von Kármán of the California Institute of

Pilots Lt Col Marion Carl and Cdr Turner Caldwell with the "Crimson Test Tube," the record-breaking Skystreak. The D-558-1's fuselage surface was made as smooth as possible by using heavy magnesium skin of 0.1in. thickness and countersunk rivets. The crimson paint was then lacquered and polished to "showroom" condition. (US Navy via Terry Panopalis)

Technology to explore the possibility of building a military aircraft that could achieve supersonic flight and beat the compressibility hazard.

At NACA's Langley Laboratory, its Director of Aeronautical Research, Dr George Lewis, noted that the British S.6B seaplane had exceeded 400mph in level flight for the first time in 1931, but he believed that aircraft using contemporary design concepts could never exceed 500mph. At that time British scientist W. F. Hilton asserted that a supersonic aircraft would require an engine of 30,000hp and he described the speed of sound as "a barrier against future progress." Undaunted, John Stack, aerodynamicist at NACA's Langley laboratory, proposed a very small aircraft with a 2,300hp Rolls-Royce piston engine and a very slender, fully streamlined fuselage, confident that it could attain 566mph.

Progress was certainly made in October 1935 at the Volta Congress on High Speeds in Aviation in Italy, which focused on supersonic flight. Delegates were shown the new Guidonia Laboratory wind tunnel, which was intended to allow testing at speeds up to Mach 2.7. Von Kármán, inspired by the presentations, asked unsuccessfully for a similar wind tunnel to be built in the USA. America thereby fell behind Germany, which constructed a Mach 4.4 tunnel at its Peenemünde research facility in 1941.

Also at the Volta Congress was German scientist Adolph Busemann, an energetic proponent of wing sweep-back, or "arrow wings." Although the concept had been employed since 1911, when John Dunne built swept-wing, tailless biplanes to improve stability, Busemann realized that wing sweep would delay the rise in drag that occurred at transonic speeds – the velocity range that extended from Mach 0.75 to Mach 1.2 – thereby increasing the speed margin at which vibration and buffeting would occur. His research paper was of great use to wartime German designers but generally ignored elsewhere. Fortunately, aerodynamicist Robert T. Jones at NACA had come to similar conclusions on wing sweep independently before 1945 while working on early unmanned, jet-powered bombs. He also favored very slender wing sections for high-speed flight.

In the absence of substantial data on supersonic flight aircraft, designers turned to the Army Ballistics Research Laboratory, where Lt Col H. Zornig had explored various shapes for bullets and projectiles and studied their drag coefficients at high Mach numbers. Ezra Kotcher of the Army Air Corps Engineering School related these findings to various aircraft profiles and was convinced that a suitably shaped aircraft could pass through the

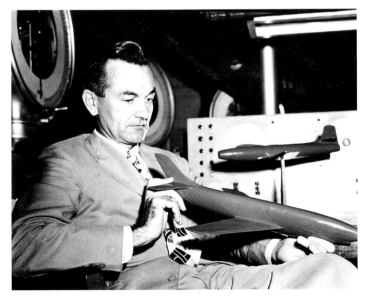

NACA's pragmatic visionary, John Stack. A proponent of high-speed research aircraft since 1933, he became a key figure in the development of the D-558 and XS-1. A D-558-1 model is posed to his left, but he is contemplating a Skyrocket-like swept-wing version that was wind-tunnel tested in December 1945. NACA recommended a straight wing for the first three D-558s, as the available power from a turbojet was insufficient to test a swept wing usefully at high transonic speeds. (NASA)

"sound barrier," since the rise in drag at transonic speed appeared to be limited to two or three times the subsonic drag value. Reporting to Gen Arnold in 1939, he advocated jet or rocket propulsion and the use of a flight research program combined with extensive full-scale wind tunnel testing. The Heinkel He 178 made the first turbojet-powered flight at that time, but it was another two years before jet propulsion was taken seriously in the USA.

America's first jet fighter, the Bell XP-59A Airacomet, used two British Power Jet W.1 engines (modified by General Electric) that Gen Arnold had requested in 1941. The fighter achieved limited production of 50 examples for service use. Its 4,000lb total thrust gave it a top speed of only 413mph and its large, straight wing and conventional structure meant that its performance offered no significant improvement on existing piston-engine types and was inferior to many. The US Navy quickly rejected it for use on aircraft carriers. It was clear that far more research was needed before a military aircraft could be developed to provide better performance. Wind tunnels capable of simulating transonic conditions were still not available in the USA and several test pilots in Britain and the USA had lost their lives trying to dive their military aircraft through violent compressibility.

In 1941 Lewis and Stack pressed for a full-scale research aircraft in the absence of suitable wind tunnel facilities. Building such types that were not prototypes of production combat aircraft was not the usual practice in the USA, but Gen Arnold supported the idea. Stack had already constructed a 500mph tunnel at Langley for propeller research, and wind-tunnel research at the laboratory had enabled the

Theodore von Kármán was a powerful influence on America's progress towards supersonic flight. He advocated high-speed wind tunnels in 1935 and promoted the idea of pure research aircraft. He was also involved in the development of jet-assisted take-off (JATO) for early, low-powered jet aircraft. It became vital for the first D-558-2 Skyrocket's take-offs. He is seen here (center) explaining JATO to Dr Clark Millikan, Dr Martin Summerfield (left), and Dr Frank Malina and Capt Homer Boushey (right). Boushey was the first US pilot to make a JATO take-off. (NASA)

When German engineer Adolph Busemann presented his paper on the benefits of swept wings for high-speed flight at the 1935 Volta Congress, its influence was not immediately apparent. However, after World War II it became clear that his ideas had been used in the design of several advanced German warplane projects. His work for NACA at Langley was vital for aircraft like the D-558-2. (NASA)

P-38 Lightning to survive near-supersonic dives by using a dive flap. From mid-1942 onwards Stack's team explored various possible configurations, including jet- or rocket-powered types. Some had swept wings and landing skids and one short-lived experiment used a Campini engine, in which a piston engine drove a compressor with a crude afterburner system.

In December 1943 the argument for turbojet propulsion in a transonic aircraft was given new impetus by Robert A. Wolf of Bell Aircraft Corporation, one of the designers of the XP-59A and a powerful advocate of the British advances in jet propulsion, which he had studied first-hand. The British had sanctioned the manufacture of a supersonic research aircraft in the form of the Miles M-52 (canceled in 1946), and it seemed appropriate for the USA also to take up that challenge. He suggested that the USAAF, USN (United States Navy) and NACA should define a suitable design, which would then be tested by NACA. Lewis felt that the aircraft should be turbojet-powered, although by January 1944, when the USAAF approved the development of a craft to investigate "aerodynamic phenomena in the range 600–650mph," it was known that Germany was well ahead in using rocket and jet power in test-bed aircraft. Kotcher, who had flown the XP-59, was also given control of the rocket-powered Northrop XP-79 flying-wing project (see X-Planes No. 10, *Northrop Flying Wings*) and its scale prototype, the MX-324, which became America's first rocket-powered aircraft in July 1944.

A subsequent comparative study of rocket versus jet power, initiated by Kotcher at the USAAF's Wright Field Aircraft Laboratory's Design Branch, concluded that rocket motors were preferable as their higher thrust offered greater transonic speed in level flight, particularly at high altitudes. He wanted a maximum speed of 800mph at 35,000ft to explore the transonic area fully. NACA, traditionally more cautious, preferred turbojets to the more hazardous and untried rocket option, particularly for an aircraft which was to take off from the ground. They ruled out the possibility of NACA pilots ever flying such an aircraft. Ironically, the first NACA pilot to be killed in a test-aircraft crash, Howard "Tick" Lilly, died in a jet-powered Douglas D-558-1. However, a tentative air-launched, rocket-driven proposal by the Design Branch featured a fuselage shape influenced by bullet ballistics studies, a faired-in cockpit and conventional but thin wing and tail surfaces. Its configuration would be reflected strongly in the USAAF-sponsored Bell XS-1 (see X-Planes No. 2, *Bell X-1*).

Transonic flight was the main topic at meetings of USAAF, USN and NACA representatives at Langley in March and May 1944. Kotcher presented the USAAF case for a rocket-driven transonic aircraft and NACA requested time to explore the possibilities, eventually offering a subsonic jet-powered type that the USAAF rejected. This initiated the link between NACA and the USN, which led to the D-558-1 subsonic project, but it was already

clear that the idea of a common research vehicle to satisfy all its sponsors was rapidly fading.

In May 1944 another useful line of research was opened up using ideas from NACA and the British Royal Aircraft Establishment on weighted aerodynamic bodies, dropped from an aircraft such as a Boeing B-29. Their behavior as they achieved supersonic speed in the drop was recorded visually and by radar, providing data that would have been more easily acquired if a suitable transonic wind tunnel had been available. Together with models attached to the noses of ground-launched 800mph rocket boosters at NACA's Wallops Island, Virginia, test site and miniature test airfoils attached to the wing of a P-51 fighter, it was possible to study the shock wave characteristics of airflow around a variety of wing cross-sections in supersonic conditions.

While Kotcher and von Kármán proceeded with further research into the Design Branch ideas, motivated by von Kármán's assurance to Brig Gen Franklin Carroll at Wright Field that a Mach 1.5 aircraft was feasible, a September 1944 memorandum from far-sighted USMC (United States Marine Corps) officer 1Lt Abraham Hyatt of the Navy's Bureau of Aeronautics Design Research Branch recommended the construction of a turbojet-powered, thin-winged pure research plane by the USN. It would explore transonic flight at a minimum of 650mph and study the behavior of the shock waves that formed around an aircraft at high speed and disrupted the airflow around it. He believed that it would assist the USN in evolving new combat aircraft. Discussions in March 1944 at the NACA headquarters in Washington, DC with USAAF and USN representatives and support from von Kármán's Board late in 1943 had already promoted that idea, although the US Navy felt the aircraft should also have military development potential.

By the end of 1944, it was clear that the USN and USAAF had such different ideas that they pursued and funded their separate research aircraft under NACA's supervision. NACA's authority, conferred by a federal charter, gave it the right to supervise projects to do with the

Bell's XS-1 rocket-powered research vehicle had a similar thin, straight wing and horizontal stabilizer to the Skystreak, and its canopy was an externally braced version of the type initially installed in the D-558-2 Skyrocket. Named *GLAMOROUS GLENNIS* after pilot Capt Chuck Yeager's wife, it was the first aircraft to break the mythical sound barrier in level flight on October 14, 1947. (USAF)

US Navy pilot Cdr Turner Caldwell, with less than 2.5 hours' experience of the D-558-1, achieved a new absolute speed record of 640.633mph in this aircraft. (US Navy via Terry Panopalis)

science of flight. Perhaps because of that, the two services entered into their respective projects with an unusual mixture of fierce competition and technical cooperation. Bell Aircraft Corporation and the Douglas Aircraft Company, the latter a major supplier of USN aircraft since its formation by Donald W. Douglas (a US Naval Academy graduate) in the 1920s, became involved in exploring suitable hardware. A small group within the Douglas design office also evolved an outline for a supersonic project that became the futuristic X-3, but that was not ordered until 1949.

No decisions on any research vehicle were finalized until the end of World War II, when two projects were approved: the USAAF-sponsored, rocket-powered Bell XS-1 and the USN's turbojet-driven Douglas D-558-1, supported by Stack's team and possibly a basis for a combat aircraft. Robert J. Woods, the Bell project engineer who secured the X-1 contract, initially favored NACA's turbojet approach and he saw the XS-1 as also having combat potential, with guns included. After Kotcher showed him information on the combat performance of the Messerschmitt Me 163 Komet rocket-powered fighter that had engaged Allied aircraft over Germany in 1944, Woods agreed that use of an Aerojet 6,000lb-thrust rocket engine could be acceptable.

The Navy's project was pushed ahead in December 1944, and requests for a new 6ft-diameter supersonic wind tunnel and a testing station at Wallops Island were added. For the USN design, NACA recommended a simple, turbojet-powered aircraft with a tubular body and an air intake in the nose. The Navy wanted side intakes so that guns could go in the nose, giving the aircraft eventual combat potential, but NACA preferred any available space to be occupied by a 500lb research instrument package rather than armament. The USN and Douglas

INSIDE THE DOUGLAS D-558-2

A cutaway view from the left side of the D-558-2

KEY

1. Instrumentation boom with yaw and pitch sensing vanes
2. Instrument and battery compartments
3. Fire extinguisher control
4. Cockpit jettison control
5. Back-rest jettison control
6. Control yoke
7. Pressure and stress measurement equipment
8. Pressure tubing disconnect manifold
9. Galvanometer and auto-potentiometer
10. Airspeed, altitude and turn meters and recorders
11. ARC-1 radio
12. Air intake duct
13. Liquid oxygen (LOX) tank, 180 US gallons
14. Expansion joint
15. Nitrogen spheres
16. Jet fuel, 145 US gallons
17. Jet fuel, 125 US gallons

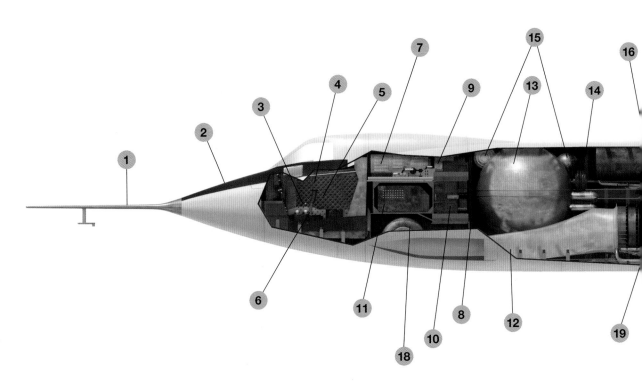

18. Nose landing gear bay
19. Westinghouse X24C (J34) turbojet engine
20. Turbo-pump system
21. Synasol tank, 200 US gallons
22. Heat exchanger
23. Turbine exhaust
24. Reaction Motors LR-8-RM-6 rocket engine
25. Engine control box
26. Nitrogen supply manifold
27. Igniter assembly
28. Rocket thrust chambers (four, 1,500lb thrust each)
29. Rocket fuel supply piping
30. Hydrogen peroxide dump pipe.

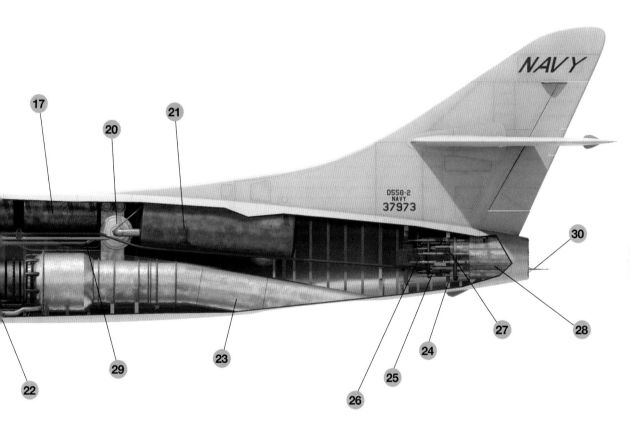

firmed up their D-558 project in February and March 1945. The basic design was meant to be modified with various wing adaptations, and six aircraft were requested so that different air intakes could also be tested. Two examples would also test rocket-assisted power and more powerful Westinghouse 24C engines to enable them to reach Mach 1. From the outset it was proposed that a second, supersonic version of the D-558 would have dual rocket and jet power and a swept wing, on the assumption that sufficient data would soon be available on using wing sweep.

The six aircraft specified in the USN's June 22, 1945 letter of intent would be flown by Douglas and NACA pilots. The total number allowed for likely casualties in this high-risk experimental area. Three aircraft would be essentially the same, one would test a single air intake in the nose, two more would have different wing airfoils and two of the six would be convertible to rocket power. The first three D-558s were all meant to explore high-speed flight up to Mach 0.89 and the three "second phase" aircraft would take the speed to Mach 1. A third phase would focus on a mock-up for a production combat prototype. D-558 wind-tunnel tests used a variety of nose- and side-mounted air intakes, but when the mock-up was built only the nose intake option remained in the design. Various wing shapes were tested, including both swept back and forward-swept types, together with a stubby, low aspect-ratio planform. Straight, thin wings were eventually specified for both the D-558 and Bell XS-1. Both designs were wind-tunnel tested with swept wings, but NACA felt that there was insufficient data on wing sweep and thought that straight wings had much better stall characteristics.

On January 27, 1946 the contract was amended to three D-558-1 aircraft with the other three being transferred to a Phase II deal as the supersonic D-558-2s. The military aspect was dropped when the Douglas Aircraft Corporation was asked to submit a design, but the required aircraft's flying characteristics and ease of handling were to be applicable to a subsequent military derivative. Douglas designers under Ed Heinemann came up with the simplest possible design combining the largest available engine, a Chevrolet-built General Electric TG-180, bench-tested in April 1944, in a small airframe. The engine was later built by Allison as the J35 with 5,000lb thrust and also used in the Republic F-84 Thunderjet fighter.

Robert J. Smith at Douglas drew up the preliminary design for the Model 558 High Speed Test Airplane in January 1945. Its minimal airframe spanned 25ft and the TG-180 with its intake duct and long tailpipe occupied much of the 35ft-long tubular fuselage. It had a tail-wheel undercarriage and fittings for three Aerojet 40AS-100 rocket booster units beneath the center fuselage. NACA's instrument package was installed ahead of the cockpit. A conventional tail unit was used with a thinner 65-008 airfoil section than the wing. The 150sq ft wing was low-mounted and given an extremely thin 10 percent thickness/chord ratio (the same as the Bell XS-1), which still allowed room for the undercarriage and 230 gallons of JP-1 fuel

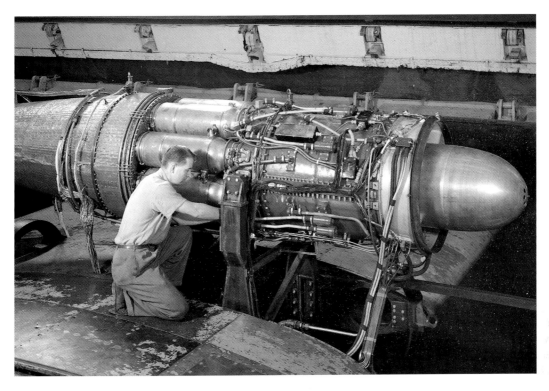

The General Electric TG-180 (J35-A-11) turbojet being tested at NACA's Lewis Flight Propulsion Laboratory in September 1947. In the D-558-1 it was mounted parallel to the wing root, with an unusually long tailpipe. It had a one-stage turbine and an 11-stage compressor, which was driven by the turbine and thereby supplied compressed air to the eight combustion chambers. The resulting hot gases then passed across the turbine wheel. (NASA via Terry Panopalis)

in its sealed forward half, with another 100 gallons planned for two wingtip tanks. NACA recommended the design and a later jet/rocket version to the USN and proposed a third stage of development in which a combat-capable aircraft would be constructed, based on the research findings with the D-558. Approval was granted quickly, although the USN's priority in April 1945 was to prove the basic concept with the first D-558-1 version before proceeding to rocket, mixed-power, or military versions.

At the same time NACA would participate in the USAAF's Bell XS-1 (later X-1). The "S" in XS-1 indicated "supersonic", which was seen to be a more unconventional project, designed specifically to achieve supersonic flight. Whereas the D-558-1 was required to take off and land under its own power, the XS-1 was designed from the outset to be launched from a carrier B-29 aircraft in order to maximize its flight time, given that its XLR-11 motor burned for less than five minutes. Several other X-planes of that time, including the Bell X-2 and early Northrop flying wings, were also designed to allow for modified wing configurations since flight testing was still the only way to compare their behavior. The design also allowed for completely detachable wings and tail surfaces so that different versions could be tested. Like the X-1, the D-558-1 had an adjustable tailplane as it was clear that small differences in its angle-of-attack could make major changes in the whole aircraft's pitch and stability at high speeds. The D-558-1's design weight was 7,500lb and it was meant to reach 625mph at sea level and Mach 0.84 at 25,000ft. Conscious that the airframe would probably have to endure considerable stress as it approached Mach 1, NACA specified that it and the XS-1 should be able to withstand 18g (positive), half as much again as any contemporary fighter aircraft.

To facilitate escape at such high speeds, an ejection seat was considered but dismissed as being too risky at that early stage in the development of these devices. Early seats used cartridges, which gave an instantaneous burst of thrust rather than the slightly more gradual, survivable boost of a rocket motor. To generate enough force to clear the tail of the D-558-1, the sudden acceleration of cartridge-powered ejection would have caused severe injury to the pilot. Instead, Douglas designed a system whereby the nose section could be uncoupled from the fuselage so that the pilot could escape backwards through its rear and parachute to safety once the nose section had slowed down sufficiently. A similar principle was used for the German Bachem Ba 349 *Natter* vertical take-off fighter in 1944 (see X-Planes No. 8, *Bachem Ba 349 Natter*) and Bell's X-2 in 1947 (see X-Planes No. *6, Bell X-2*). Heinemann estimated that it would be safe at speeds below 350mph, a much lower limit than Bell estimated for their escape procedure. For the D-558-1 it was the only means of escape, as the forward-opening cockpit canopy was locked down in flight.

While construction of the D-558-1 mock-up proceeded, two senior Douglas designers, L. Eugene Root and A. M. O. Smith, returned from visiting Germany in May 1945 with data that would enable design of their second-generation D-558s: the swept-wing, supersonic D-558-2. Root had been inspired by Cdr Emerson Conlon at the USN's Bureau of Aeronautics to examine Hyatt's proposal for a transonic research aircraft and instigate a detailed design. Only flight-testing of the D-558-1 could show whether it would actually survive transonic flight.

Confidence in that venture would be temporarily shaken on September 27, 1946 when Geoffrey de Havilland, flying his company's DH 108 Swallow that was built to test swept wings, attempted to beat a world speed record of 615.77mph set three weeks previously by an RAF Meteor F 4. Its pilot, Grp Capt Edward "Teddy" Donaldson, was a Battle of Britain Hurricane ace who commanded the RAF's first operational Meteor jet squadron. Entering severe compressibility at near-sonic speed over the Thames Estuary, the Swallow was torn apart, killing de Havilland and reinforcing theories about the impossibility of breaking the "sound barrier." On September 6, 1948 test pilot John Derry did see Mach 1.04 on the Mach meter (later confirmed at Mach 1.02) of DH 108 VW120 as he struggled to recover from a near-vertical dive.

Meanwhile, the world speed record had been recaptured by Col Albert Boyd on June 17, 1947 in a specially modified Lockheed XP-80R, although only by an 8mph increase on the British record. Within two months a D-558-1 would increase that margin significantly, although there was an understandable degree of reluctance by many pilots to fly the aircraft or its rocket-powered successor, the D-558-2 Skyrocket, in what was clearly a very risky endeavor.

CHAPTER TWO

THE CRIMSON TEST TUBE

Gene May lifts the first Skystreak briefly into the air on April 21, 1947. A drop in fuel pressure reduced the flight to less than a minute, a repeat of the first attempt at flight on April 15. The aircraft has the original bubble canopy. (US Navy via Terry Panopalis)

In the first week of July 1945, the D-558 mock-up was completed and inspected. Its air intake was now a simple bifurcated, circular version in the nose rather than the side-mounted type, which NACA considered less well-researched. During the inspection, Stack expressed reservations about the strength of the bubble-type cockpit canopy (particularly if the aircraft rolled inverted on landing) and also the extent of the pilot's visibility through it.

There were doubts about the extremely cramped space for the pilot in a small cockpit, 22 inches wide, wedged between two air intake ducts. To read the dials at the top of his instrument panel a pilot had to hunch down in his seat, restricting his forward vision. A test pilot, Stan Butchart, reported, "You flew it with your elbows in and the wheel between your knees, and crunched down. Your head was up into a tight canopy." If a pilot wearing a helmet turned to look out sideways, his helmet would get stuck and he would have to lower himself sufficiently to turn and look to the front. Butchart even had difficulty in reaching down between the control column and his emergency oxygen bottle to operate the landing gear handle, an excuse for some ribbing by his chase pilot, Capt Chuck Yeager. NACA also criticized the minimal space available for their research instruments. More room was allowed for them in the D-558-2.

The main structural complexity was found in the cockpit section separation mechanism. Heinemann engineered a system using four retracting hooks (similar to bomb-bay shackles) that released the nose cleanly when the pilot pulled a handle above the instrument panel. His team also devised a way of disconnecting the various electrical

cables running between the nose and main fuselage by using pull-out plugs. The mechanical flight surface control rods were more difficult to separate, but engineers used a system with a frangible shaft that ran between two drums, one each side of a bulkhead that marked the separation point. Moving in unison, the drums translated the push-pull motion of the control rods into a rotary movement for the part of their travel that crossed the nose separation point.

One-fourteenth scale models of the nose section were tested in the Langley Laboratory's 20ft free-spinning wind tunnel, together with similar models for the D-558-2. The researchers reported that:

"When the model with no [stabilizing] fins was launched it usually went immediately into a tumbling motion. Assuming that the rate of tumbling varies directly with the speed at which the nose section travels through the air and that the nose section was jettisoned from the airplane at high speed and went immediately into a tumbling motion, it is probable that much higher acceleration forces would result, which would be very dangerous to a man in the pilot's seat."

Forces of up to 15g acting on the pilot's head were anticipated.

NACA found that stabilizing fins placed near the base of the nose section "prevented the model from tumbling and made it descend in a stable, nose-down attitude." The addition of a braking parachute to stabilize and slow the nose immediately after launch raised many problems. "The parachute may fail to stabilize the nose section until after dangerous tumbling has taken place. A severe parachute shock load would develop and there was a possibility of fouling of the parachute on the rest of the airplane before it could open fully."

Douglas could not obtain a suitable brake parachute to slow the nose section more safely, and the extending stabilizing fins were rejected to

Gene May pilots D-558-1 BuNo 37970, with minimal markings, on October 9, 1947 during the high-speed test program. May was testing some re-work on the aircraft after its world speed record flights two months previously. (US Navy via Terry Panopalis)

Displayed in sections, the second D-558-1 is seen with the first example and a pair of wingtip fuel tanks that were used later in the program. The narrow cockpit area, with air inlets to each side of it and a very small windshield, can be seen in the jettisonable forward fuselage section. (Terry Panopalis collection)

save weight and space on the assumption that aerodynamic braking would eventually slow the object sufficiently for a safe bail-out from its rear. However, it was clear that the jettisonable nose was unlikely to be a safe way to leave the aircraft, although it was arguably an improvement on the escape provisions for XS-1 pilots. They had no means of escape other than jumping from a side hatch and almost certainly hitting the wing or tail of the aircraft in so doing. For them, landing the aircraft somehow was the only practicable option. For D-558 pilots the situation was actually much the same, and test pilot Bob Champine was not alone in believing that escape was impossible, and that the only hope was "just to land the thing."

Further airframe refinements were prompted by NACA recommendations before the first D-558-1 was constructed. The horizontal stabilizer was mounted higher on the vertical tail to avoid turbulence from the wing and its airfoil section was made thinner than the wing's airfoil section as NACA suggested. Work on the XS-1 confirmed that the angle-of-attack of the horizontal stabilizer should also be made adjustable in flight. It was actuated electrically by a switch on the pilot's control wheel. The flying controls were otherwise manually operated without hydraulic boost or trim tabs, and the pilot was given a control wheel rather than a stick in case the effort of operating the controls required both hands. Bell used the same arrangement in the XS-1.

Hydraulic dampers were installed to counteract flutter and speed brakes were installed at the final stages of the design process at the request of pilots. The tailplane surfaces were stiffened internally. The entire airframe was extremely strong and the flying surfaces were rigid with comparatively thick magnesium alloy skins over aluminum frames, dispensing with the need for many internal stiffeners. Test pilot Scott Crossfield recalled that "there was rudder buzz at about Mach 0.999. That was the only dynamic problem I remember on any of those airplanes." Its JP-1 fuel load weighed only 1,400lb compared with the 5,000lb of rocket fuel in three different supply systems that the smaller XS-1 had to carry.

NACA's crucial instrument package, supervised like most of the X-plane instrumentation by Gerald Truszynski, and the cockpit were supplied with cooled air. Four hundred small holes or orifices were opened in the wing and tail surfaces for pressure distribution sensors to be installed and linked to recording manometers. The installation

The narrow confines of the D-558-1 cockpit with conventional flight instruments plus a Mach meter that goes just beyond Mach 1. Placards warn the pilot not to lower his landing gear above 300mph or exceed 160mph landing speed. Pilots preferred fast landings, touching down at 143mph, as Skystreaks tended to roll suddenly from side to side as they approached stalling speed, which was relatively high due to the small wing area. (NASA)

of the orifices and their ducting was a formidable design challenge in itself. Other control forces, engine performance data, and flight conditions were recorded on a 30-channel Miller oscillograph.

Making the main undercarriage retractable into the thin wings posed considerable problems. The fuselage interior was too crowded to accommodate landing gear, so the Goodrich Corporation and Bendix were asked to develop slender 20 x 4.4-in. wheels and tires (inflated to 230psi) for the undercarriage components. These were to become a handicap in the early flight test program.

TESTING THE "TUBE"

Construction of the first aircraft (BuNo 37970), named Skystreak, began in March 1946, and it was ready for delivery to Muroc Dry

D-558-1 BuNo 37970

Test pilot Gene May flew the first Skystreak (BuNo 37970) on its 93rd contractor's flight on November 4, 1948 to test the installation of 50-gallon wingtip fuel tanks at speeds in a dive up to Mach 0.945. Endplates were attached to the tanks for these high-speed dives.

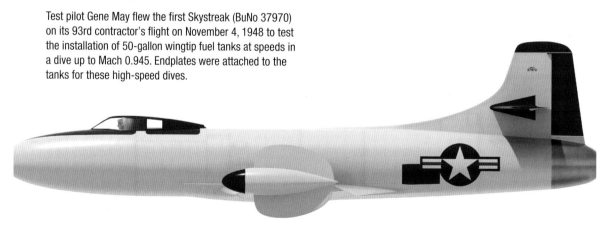

Lt Col Marion Carl leaves the first D-558-1 after his record-breaking flight on August 25, 1947. The small, reinforced high-speed canopy, locked internally with two handles, was installed on the first Skystreak after its 12th flight to improve pilot protection if the aircraft rolled over on landing. A pilot's helmet could easily get stuck within its narrow "V" shape. It was such a tight fit around the pilot's head that he had either to wear a soft leather helmet or cover a hard helmet with a chamois leather cloth to prevent it from marking the glass canopy panels. Test pilot Stan Butchart described it as the ideal aircraft in which to get claustrophobia. (US Navy)

Lake (later renamed Edwards AFB) on April 10, 1947. At that time construction of the second generation D-558-2 was already underway at Douglas Aircraft's El Segundo plant, and it was completed in November 1947 and at Muroc by December 10. Meanwhile, initial flight testing of the Skystreak began by Douglas company pilots. The first of these was 42-year-old project pilot Eugene F. May, who had tested many Douglas aircraft, including the A-26 Invader and snub-nosed B-18 Bolo twin-engined medium bombers.

Fifteen months after the XS-1 made its first unpowered flight, May attempted a first Skystreak flight on April 15, 1947 but, after slow acceleration and a long ground run, the engine faltered on take-off. Fuel pressure suddenly dropped, reducing engine thrust and forcing him to land immediately. The left brake unit fell apart as he slowed. A second attempt on April 21 ended the same way after less than one minute airborne. The next flight on May 29 ended after seven minutes when a fire warning light illuminated and the main landing gear refused to retract. The engineers were baffled, as the fuel pressure fault could not be replicated in ground tests. Enlarging the vents for the fuel tanks in the wings made some improvement, but similar problems plagued the next six flights. The July 3 attempt was the first in which worthwhile testing was possible after additions to the hydraulic landing gear retraction mechanisms. May was able

to perform two stalls during the 19-minute hop. Flight time was increased to 33 minutes ten days later, and he was able to take the aircraft to 30,000ft and Mach 0.75.

By July 13, Douglas engineers were satisfied with the aircraft's low-speed handling and they prepared for more demanding flights. The light-weight bubble canopy was replaced with a reinforced high-speed version with a V-shaped windshield, and on August 3 the speed was pushed to Mach 0.85 at 30,000ft, which May thought was the "practical safe limit at high altitude." It was also faster than the XS-1's best performance at that time, but Col Klein of the USAAF's Fighter Branch Projects Section stated that the XS-1 was also intended to be a transonic research vehicle and assured his audience of engineers that reaching supersonic speed was a priority for later models than the XS-1.

A tailpipe extension was tested briefly on the first two Skystreaks to improve cooling in the rear fuselage. The modification was removed after cracking occurred in the metal. (US Navy via Terry Panopalis collection)

The first Skystreak was fitted with endplates attached to its tiptanks as a way of preventing airflow from moving span-wise at high speeds. It was dived to Mach 0.945 in this configuration on November 4, 1948. (NASA via Terry Panopalis)

Lt Col Marion Carl, Gene May, and Cdr Turner Caldwell who, with Larry Peyton, made the majority of the 129 contractor's flights in the three Skystreaks. (US Navy)

His views were certainly at odds with many involved in the program, who believed that the little orange plane was intended primarily to "break the sound barrier."

In August the second D-558-1 (BuNo 37971), without instrumentation, joined the test program and after initial contractor flights by May it was flown by two service pilots, Cdr Turner F. Caldwell, Jr and Maj Marion Carl. Caldwell became the USN's project officer for the D-558 and Carl, a Pacific War ace, had experience of several early jets including a captured Messerschmitt Me 262. He also became the first USMC pilot to qualify on helicopters. At 6ft 2in in height, he found the Skystreak's cockpit extremely cramped, and the tiny canopy made it impossible for him to wear a hard flying helmet.

At a time when world speed records were a subject for strong inter-service and international competition, the possibility of using a D-558-1 to beat Donaldson's British record of 615.77mph arose. The idea persisted after Boyd raised the record speed in his P-80R and the USN saw such an attempt as a good way of allowing service pilots to show the Skystreak's favorable handling qualities at low altitude and high speed.

RED RECORDS

Both pilots made Skystreak familiarization flights on August 18. While the handling characteristics were smooth and pleasant at lower speeds, the aircraft began to develop oscillation and buffeting above Mach 0.75. Two days later Cdr Caldwell flew BuNo 37970 over a 3km course at altitudes below 75m, averaging 640.7mph (Mach 0.825) during four passes. His fastest run hit 653.372mph. Maj Carl also flew that day but reached a lower speed and asked for the TG-180 to be tweaked to give 102 percent power in flight. On August 25 Carl made a second attempt, reaching an average of 650.796mph (Mach 0.828) with Bob

CALDWELL SETS A NEW WORLD SPEED RECORD

NEXT PAGES

On August 20, 1947 Cdr Taylor Caldwell, the USN's D-558-1 Project Officer, flew the first Skystreak at an average speed of 640.663mph, setting a new world speed record. During the flight he sustained an altitude of 75ft above the dry lake bed at Muroc and beat the previous world speed record of 616mph set by Col Albert Boyd. Caldwell's record was broken five days later by Lt Col Marion Carl, who reached 650.606mph in the same aircraft.

Champine as chase pilot. Although May apparently flew at a new record speed of 680mph later in 1947, the glory returned to the newly created USAF when it was revealed that Capt Chuck Yeager had flown the XS-1 to Mach 1.06 on October 14, 1947, scoring the world's first supersonic speed record.

The personal rivalry between May and Yeager, a reflection of the overall inter-service competition at the time, was for the moment settled in the USAF's favor, although it persisted in other ways. Although Yeager believed the competition was good for aviation and research, his animosity towards May was more personal. May still resented the occasion on which Yeager, "real annoyed" at USN comments that the XS-1 was not a genuine supersonic airplane because it was air-launched rather than taking off normally, had upstaged his Skystreak demonstration for USN top brass by making a near-sonic pass in front of the audience just as the Skystreak passed at appreciably slower speed. Yeager riposted to the "air-launch" slurs on January 5, 1949 at the suggestion of Larry Bell (XS-1 manufacturer) by blasting off from a runway with a half-load of fuel in his XS-1. He made a brief 1,500ft take-off run, became airborne at 200mph, and climbed to 23,000ft at Mach 1.03 80 seconds later. No one doubted that he knew of the USN's intention to use the D-558-2 to make the first rocket-only take-off. Although his display was of little use as a research flight, Yeager had made his point and won bragging rights once again, particularly when the USN soon shared the XS-1's usual air-launch method for their D-558-2.

A low pass with a long "streak" of black smoke during the record attempt flights in August 1947. Capt Frederick Trapnell found the D-558-1's lateral control "undesirably coarse" because of the small control wheel. However, like most of its pilots, he praised its flying characteristics generally and its high standards of engineering and design. The very small cockpit restricted the space for all the piloting activities. A control wheel was chosen rather than a conventional stick to provide more control force for the pilot under buffeting at high speeds. (US Navy)

THE NACA DEAL

The first D-558-1 was returned to Douglas's contractor testing program with May as pilot, while the second example continued testing with a

In white finish and the markings of NACA, the third D-558-1 (BuNo 37972) allows its groundcrew a break. After a single clearance flight by Gene May it was handed over to NACA on January 22, 1949. The jettisonable portion of the nose separated along a vertical line behind the canopy, but it could only be used to escape within a very narrow range of speed and altitude parameters. (NASA via Terry Panopalis)

variety of pilots including Caldwell, Carl, May, and USN pilot Capt Frederick Trapnell, together with Douglas pilots George Jansen, Johnny Martin, and Larry Peyton. May's program included a number of flights from October 24, 1947 to November 1948 with 50-gallon tiptanks fitted, including dives to Mach 0.96 and jettisoning of full tanks. The tanks helped with directional stability at high speeds by preventing the airflow from separating span-wise and escaping off the ends of the wings, thereby interfering with its lift characteristics. To increase their effect on straightening the airflow direction, Douglas added thin vertical endplates to the tanks, and Gene May flew the aircraft to Mach 0.948 on November 4, 1948 with the endplates installed.

He also put the Skystreak through a series of tests in dives from high altitude, reaching speeds above Mach 0.94 and requiring considerable effort on his part to pull out under forces of 3g. On September 29, 1948 he entered a 35-degree dive and the D-558-1, for once only, went supersonic as it briefly approached an out-of-control situation. In all, Gene May made 121 flights in the aircraft, more than half the total for all three Skystreaks.

Of that trio the second aircraft, which he flew only seven times, had the shortest life. After 26 flights by Douglas and USN pilots it was passed to NACA in October 1947 and painted white since NACA's Walt Williams, heading the Muroc Flight Test unit, thought that it would become more easily visible for optical tracking against a deep blue sky on test flights than the crimson D-558-1. All future maintenance and alterations to the second and third aircraft were

The third D-558-1 Skystreak, BuNo 37972/NACA 142 after modifications following its first four flights and the loss of the second aircraft in May 1948. Test pilot John Griffiths from NACA's Cleveland Laboratory made the fifth flight and joined Bob Champine to continue the Phase 1 test program. The aircraft made seventy-eight NACA flights, with eight different pilots, the last being Scott Crossfield on October 6, 1953 to explore low speed stability. All the pilots who flew it were impressed by the Skystreak's handling qualities, although it became a challenge in high-speed, near-supersonic dives.

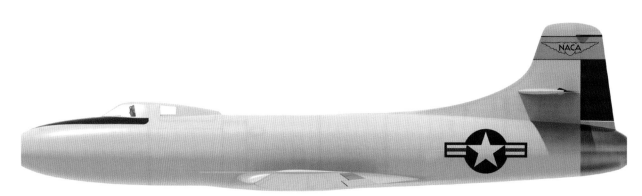

X PLANES

SKYSTREAK BuNo 37972 (NACA 142)

to be undertaken by Douglas and paid for by the USN. Engine maintenance or modernization would be similarly funded, while the USAF paid for fuel. It was a clear and practical division of responsibility.

NACA installed its full 634lb of research instrumentation in the second aircraft, using a package that was similar to the one used in the Bell XS-1 and X-2. All control surfaces were equipped with sensors, performance and control forces were recorded and a small gunsight camera was installed to monitor the readings on the pilot's instrument panels. A similar camera would provide the most useful information on the disastrous loss of Bell X-2 46-674 and its pilot after exceeding Mach 3 in September 1956.

On November 25, 1947 Howard Lilly began a series of 19 test flights to explore stability, control, and aerodynamic loads. West Virginian "Tick" Lilly was an experienced NACA pilot who had flown the XS-1 and became the third man to "go supersonic" in it. Like several other test pilots, he had also been successful in flying carefully tuned piston-engined fighters in the Thompson Trophy Race. In the D-558-1, he was required to test its directional stability with a series of side-slips, "rudder kicks," snap rolls and pull-ups across the full range of speeds at altitudes between 10,000ft and 30,000ft. On seven of these flights landing gear malfunctions occurred, three were curtailed by instrumentation problems and on one a burned-out inverter filled the cockpit with smoke, ending the flight.

The May 3, 1948 flight was abandoned because the undercarriage would not retract and lock, but Lilly took off again that afternoon, following repairs to the door latches. He had climbed to 150ft at 250mph, only two miles from take-off, when the early J35-C-3 engine's compressor disintegrated. Its blade fragments blasted skin panels from the fuselage, severed the rudder and elevator cables and fuel lines and

D-558-1 37972 shows off its unusual landing gear spacing and early markings for the NACA in October 1949. The D-558 series aircraft were once described as "the X-planes that weren't." Although their purpose was similar to that of the Bell X-1, X-2, and others, the "X" prefix was a USAAF-only idea, and it did not apply to joint-service research aircraft projects until the North American X-15 was developed. (NASA via Terry Panopalis)

started a fire. Lilly was far too low to use the ejection system and the aircraft suddenly rolled to the left, hit the ground inverted and burst into flames. The confining cockpit canopy did not allow space for Lilly to wear a protective helmet, but the disintegration of the cockpit area sealed his fate in any case. It was the first serious accident in an otherwise very successful program of 127 flights to date, and it caused a lengthy hiatus in the flying tests.

Among the recommendations of the Board of Investigation following the fatal accident was a proposal to redesign the canopy as well as providing the compressor with some armor-plating to contain an explosion. More oddly, it proposed that the nose separation system should be flight tested in a surplus, autopilot-controlled D-558-1 *after* the completion of the test program. Interestingly, it also suggested that the air-launching method from a carrier aircraft used for the XS-1 and rejected by the USN would be a "viable means of launching research aircraft" and it would "give the pilot more reaction time in case of an emergency."

Douglas did modify the canopies of the two surviving aircraft and armor-plated their emergency fuel pumps and fuel lines with 6.35mm stainless steel protection where they passed through the engine bay. The forward-opening canopy was instead hinged at the rear to allow some chance of escape if the cockpit jettison method was inoperative. They also added a back-up set of control cables and new fuel hoses with wire-wound protective covering. The extra weight penalty had to be accepted, although it was decided to retain the original red paint on the flight-control surfaces of the third aircraft when it was given the white NACA scheme in case the weight of the paint upset their balance and caused flutter.

With the modifications in place, NACA inspected and accepted the third aircraft (BuNo 37972) as "NACA 142" on January 22, 1949 to replace the ill-fated second Skystreak. NACA also took on the redundant first aircraft as a spares source when it finished its 101-flight program. The jet's instrumented horizontal stabilizer was borrowed in June 1950 for NACA 142 to test tail loads, elevator vibration, and the effects of vortex generators above the wings in improving overall stability.

A safe landing at Muroc by the second Skystreak. On its 19th NACA flight on May 3, 1948 it crashed on take-off, killing test pilot Howard Lilly. Not until then were concerns expressed about the restricted access to the engine for inspection purposes. The compressor failure that caused that crash might thereby have been anticipated. Removal of the engine was difficult, involving the disconnection of all flight instrumentation as well. Only one main engine access panel was provided. During modifications after that accident the fuel and control lines were given protection. (NASA)

EUGENE FRANCIS MAY (1904–66)

Born in Shelby, Alabama, to a silent movie theater impresario, Gene May made an early decision to become a pilot after seeing his first aircraft at a county fair as a child. He joined the US Navy, hoping to learn to fly, but was made a radio operator on board the battleship USS *West Virginia* (BB-48). Returning to his young wife he began self-funded flying lessons in 1926, supporting his family with manual jobs and eventually moving to Detroit, where he received his pilot's license. After a few years as a display pilot and flying instructor May resolved to become a test pilot and began intensive study of aeronautical engineering, funding himself through a traveling salesman job.

In 1937 he became a TWA captain flying DC-3s, moving to Canada when World War II began to become an instructor for RCAF crews ferrying bombers to Britain. By 1941 May decided to return to the USA and was offered a job with Douglas Aircraft, where he conducted production check flights on A-20 light bombers and other company-produced warplanes. He earned a reputation as a strong-willed individual, unafraid to speak his mind. Although he was already a 42-year-old grandfather, he became the D-558-1 project pilot in 1947 and expected to be the first man to break the "sound barrier" in it. His highly respected understanding of practical aeronautical engineering made him the ideal pilot to cope with the aircraft's early in-flight problems and he officially logged at least 99 of the D-558-1's 231 flights and 37 more in the D-558-2.

His last test flight, on December 1, 1949 was a Press demonstration in the first Skyrocket that began with a jet-powered pass at 50ft. May followed up with an ear-splitting pass with all four rocket chambers burning, reportedly at supersonic speed, followed by a vertical zoom climb. He turned down the chance of piloting the air-launch phase of the D-558-2 program and became Head of Flight Operations for Douglas, overseeing the D-558-2 underwing stores tests and working on the DC-7 and missiles. Moving to Tulsa, Oklahoma, where Douglas was to build B-47 Stratojets for Boeing, he soon left the company and in 1952 became a corporate pilot until his retirement in 1960 and death from heart disease in 1966.

Among the eight NACA pilots who flew D-558-1s were Stan Butchart, John McKay, Joe Walker, Robert Champine, and A. Scott Crossfield, all of whom would play major roles in the development of a number of X-planes including the X-15 (see X-Planes No. 3, *North American X-15*). John H. Griffith also joined the team. Flights with the third aircraft re-commenced on October 31, 1949 and continued for seven months. Griffith commented that the Skystreak was:

"A beautiful airplane and really a lot of fun to fly. We were doing a lot of flying between Mach 0.8 and Mach 1.0. Quite a few of the flights that I was on were very close to Mach 1. We were measuring the pitch characteristics and, of course, the pressure distribution over the wing and all the stability and control aspects of flying through Mach 0.80 to 0.99, which was giving us a lot of information that was pretty much needed to keep these airplanes out of trouble when they got going in that speed range. The wing [area] on the D-558-1 was about 150sq ft and that made the stall speed a little high in some cases. I know that I was doing a clean configuration stall…and I felt pretty good at 150kts IAS. At about 149kts I dropped 1,000ft. And so things quit all of a sudden. The aircraft simply stopped flying."

Crossfield and Butchart took on the tail-load flights with others to study the effectiveness of ailerons at high speeds. McKay, who would be seriously injured in an X-15 mishap, flew additional tests with the wingtip tanks to give prolonged flight times, and Crossfield made a low-speed stability test on June 10, 1953 which proved to be the last of 83 flights for NACA 142. Former Naval Aviator Bob Champine was persuaded by Herb Hoover at NACA to take over Lilly's role as

Bob Champine, removing his parachute in this image, replaced Howard Lilly as a D-558-1 test pilot for NACA from April 1949, making a series of high-speed dives in NACA 142 to explore its lateral stability. He landed after one flight with only 11 gallons of fuel aboard. After Lilly's accident, the aircraft's canopy hinge was moved to its rear edge. (NASA)

project pilot. He had flown the XS-1 13 times and also the Bell L-39, a swept-wing version of the Airacobra that provided some much-needed practical experience of wing sweep for the company's X-2 rocket plane. The Skystreak was pushed to its limiting speed of Mach 0.99 in a mid-June 1950 flight by Griffith, who had also flown the XS-1, at 37,000ft.

STABILITY STUDIES

More than 30 of the Skystreak's flights in 1950 explored tail loads and longitudinal stability, subjects of constant concern to all X-plane researchers as they tried to establish why one wing of the aircraft would suddenly dip downwards at transonic speeds, followed by an uncommanded roll when the pilot tried to use the ailerons to correct the motion. This tendency was usually accompanied by a loss of elevator effectiveness and unintended trim changes as speed increased beyond Mach 0.85, together with a six-fold increase in the force required to trim the aircraft using the control wheel.

The onboard instruments showed that a shock wave that developed over the center of the wing chord at about Mach 0.76 steadily moved aft as speed increased. At Mach 0.82 a second shock wave formed below the wing and moved further aft than the first shock wave, shifting the aircraft's aerodynamic center of pressure aft and causing a nose-down attitude. This movement continued as speed increased further so that that the wings and tailplane no longer provided their normal lift and stabilizing properties. Vortex generators were fitted to the wings of the D-558-1 and XS-1 and they had some beneficial effects in controlling the airflow and preventing wing drop above Mach 0.89.

In December 1950 the first of 29 flights was made by Crossfield to explore the aircraft's stability when it encountered buffeting caused by turbulent airflow. In early test flights Douglas had recorded such serious vibration in the elevators that it had prescribed an arbitrary speed limit of Mach 0.92 on the D-558-1. NACA's Muroc research station was concerned that the elevator vibrations occurring at that speed had not been properly corrected by the Douglas Company before the aircraft was handed over to NACA, apart from some "shot in the dark" changes to the elevator balance weights. As a precaution

the elevators were replaced in May 1951, although no evidence of impending fatigue failure had been noticed at that stage. The Skystreak was also fitted with a nose-boom incorporating an airspeed sensor and angle-of-attack vane.

This research continued until late 1951, and after some serious refurbishment the Skystreak began a final series of flights to gather more data on lateral stability and control for use in the programs for the second-generation D-558-2. The last of these was made by John McKay on April 2, 1953.

In December 1952 Donald Bellman at NACA reported that analysis of tail load and tail buffeting was "dormant because the engineers have been needed for work on the D-558-2 and X-3 programs." NACA ordered six more flights up to June 3 to test buffeting with the tiptanks in place. Scott Crossfield made the last of 225 Skystreak flights on June 10 and the aircraft was then stripped of its research instrumentation, used as a test pilot trainer and then returned to storage like the first example after a long and useful active life. The performance of serving military jets had long since overtaken the Skystreak's and the newer aircraft became ready sources of data on high subsonic flight, making the D-558-1 surplus to requirements.

D-558-1 BuNo 37972 over the Edwards test range in May 1952, with an instrumentation boom attached to its nose. When NACA Muroc's Engineer-in-Charge, Walt Williams, inspected the D-558-1 in October 1946, he was disappointed at the slow progress compared with the Bell XS-1 but impressed that it was "based on more sound engineering than the XS-1. We are getting a better research vehicle even though it does not have the speed potential." (NASA)

DOUGLAS D-558-1 TECHNICAL SPECIFICATIONS	
Wingspan	25ft, with NACA 65-110 airfoil section
Wing area	150.7sq ft
Length	35ft 8.5in.
Height	12ft 1.7in.
Power	One Allison J35-A-11 (General Electric TG-180) turbojet rated at 5,000lb static thrust
Fuel	230 gallons plus 100 gallons in optional wingtip drop tanks
Weights	take-off 10,105lb, landing weight 7,711lb
Max speed	652mph at sea level (Mach 0.832), reduced by 10mph with tip tanks
Range	490 miles (856 with tip tanks)
Ceiling	45,600ft
Rate of climb	9,220ft/min

SKYROCKET

The Skyrocket's bullet-like fuselage shape was at first enhanced by the flush canopy and smaller vertical stabilizer. Its tapered tail cone was a temporary cover while the LR-8-RM-5 rocket motor was awaited. When the first Skyrocket was rolled out, the USAF version of the LR-8 had already accumulated a year of flight time in the Bell XS-1. (US Navy via Terry Panopalis)

Following Gene May's 680mph flight in a D-558-1, announced in November 1947, it was assumed that an attempt at supersonic flight would follow shortly. In fact, Yeager had already achieved that goal on October 14 and on most of his seven later flights in the XS-1, but his achievement was subject to tight security. North American test pilot George Welch had also reached supersonic speed in the prototype XP-86 Sabre, soon to become a production jet fighter, on April 26, 1948. The press broke the news of Yeager's record unofficially in December but it was not confirmed by the USAF until June 15, 1948.

By that time the USN's new Mach challenger and dedicated research "flying laboratory," the Skyrocket, had been publicly revealed for six months after the signing of a revised contract on January 27, 1947 that canceled the final three Skystreaks and replaced them with three more powerful D-558-2 Skyrockets. They received the BuNos 37973 to 37975, which had already been allocated for the second Skystreak trio.

Bell originally planned to give its XS-1 a swept wing, but the company responded to the lack of hard data on wing sweep and used a straight wing. When the later X-2 was conceived, it had a swept wing and tail, but it was also virtually a new aircraft. In conceiving its D-558 project, Douglas also effected a major redesign. Following Stack and Kotcher's recommendation that swept wings should be used only on the second batch of three aircraft, the "Phase II" or D-558-2 adaptations of the same basic aircraft had wings swept at 35 degrees and tailplane sweep of 49 degrees.

To test the German data on swept wings recovered and microfilmed by Root and Smith from Germany's Göttingen research laboratories

in the summer of 1945, the Bell designer Stan Smith arranged for the USN to use two P-63C Kingcobras, renamed with the USN designations L-39-1 and L-39-2 (or XP-63Ns). Their normal straight wings were replaced with wing panels taken from surplus P-39E fighters and swept at 35 degrees, or 40 degrees for L-39-2. Adjustable leading-edge slats, wing fences, and flaps were installed, as the main purpose was to explore the wing's behavior at low speeds. The L-39-1 was dedicated to research for the D-558-2, while the second example was given a sharp wooden leading edge to simulate the bi-convex wing section that was planned for Bell's own X-2. NACA test pilot John Griffith found that it was "the first airplane I ever flew in which you could push on the left rudder and the airplane would roll right."

Although development of the Phase II aircraft overlapped that of the D-558-1 and XS-1, the evolution of its swept wing profile took almost a year. After extensive tests in the Langley high-speed wind tunnels, NACA recommended a thin, 10 percent chord airfoil. The aircraft was supposed to have the same low-speed characteristics as the straight-winged version and Douglas had responsibility for the low-speed and stalling tests. Duplicating the Skystreak's relatively low stall speed meant enlarging the D-558-2 wing area by 25sq ft to 175sq ft. It was known that poor low-speed handling could be a disadvantage of swept wings, but Douglas wind-tunnel tests at the California Institute of Technology were reassuring in that respect. The horizontal tail surfaces were given 40 degrees of sweep so that they would reach their critical Mach number and the onset of buffeting at a different speed from the wing.

NACA approved the use of a USN-sponsored Reaction Motors Inc. LR-8-RM-6 (A6000C4) four-cylinder rocket motor to compensate for the lack of more powerful turbojets. A modified version, the XLR-11 was used in the Bell XS-1 and in the early stages of flight trials of the

The D-558-2 in its November 10, 1947 roll-out configuration. The one-piece flush canopy and short vertical tail were soon replaced. Although they created minimal drag, the flush-mounted air intakes, with a variable position lip, contributed to the lack of airflow through the turbojet engine, which was in any case severely deficient in thrust. The downward-canted jet pipe also reduced thrust. (US Navy)

RIGHT
P2B-1S NASA 137 is here raised on hydralic maintenance jacks to allow a D-558-2 to be suspended in its belly. The extensive modifications and external reinforcements that the bomber has undergone in order to carry the rocket aircraft are evident here. (NASA via Terry Panopalis)

ABOVE
Gene May makes a high-speed run in
the first D-558-2 on November 29,
1949 after its rocket motor had been
installed. His 21-minute flight was the
first with both jet and rocket engines
at full thrust. (US Navy)

North American X-15. Developing 6,000lb of thrust, the relatively
small, compact engine was installed in the tail, but the separate ethyl
alcohol with water mixture and liquid oxygen (LOX) fuel supplies
took up a considerable amount of space and the motor burned the
propellants at 2,000lb per minute. Each of the four rocket chambers
developed 1,500lb of thrust. If the aircraft was to take off from the

The third D-558-2 (BuNo 37975, which later became NACA 145) was the first Skyrocket to reach supersonic speed, a feat achieved on its eleventh contractor flight on June 24, 1949. Pilot Gene May described the flight as "glassy smooth". It had an LR-8-RM-6 rocket motor and a J34 turbojet unlike the "pure rocket" second Skyrocket. Its first fifteen flights began with take-offs from the ground, but on September 8, 1950 it made the Skyrocket's first air-launch. Bill Bridgeman used the jet engine for that flight but air-launched rocket-motor flights began on November 17. Its first NACA jet and rocket air-launch was not until May 17, 1951 and on that occasion the turbojet flamed out.

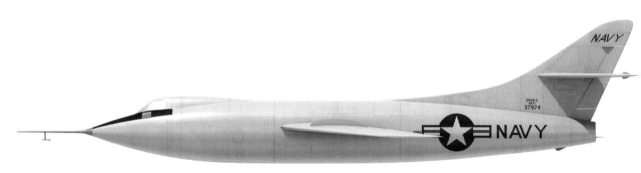

X PLANES
D-558-2 SKYROCKET BuNo 37975

ground like the Skystreak, the flight time with rocket power only would be severely limited and virtually useless for research purposes.

While the USAAF avoided this problem by air-dropping their XS-1 from a modified B-29 Superfortress at more than 30,000ft, the USN initially shunned this solution to reduce cost and simplify the flight process. Instead, a second power source in the form of a Westinghouse 24C (J34) turbojet was also installed, with 250 gallons of jet fuel in two center-fuselage tanks, for take-off and return to base after a brief 1 minute 40 seconds dash with the rocket motor at its full 6,000lb thrust, or 2 minutes at a reduced 4,500lb power. The J34 axial-flow turbojet was mounted in the central fuselage area, drawing air from two low-mounted intakes ahead of the wing. It had an 11-stage compressor, a double annular combustion chamber, and a two-stage turbine. Its 3,000lb thrust passed through a long jet-pipe and exited at an 8-degree downward angle beneath the rear fuselage. In practice, the engine was barely able to get the aircraft airborne without rocket assistance, and take-off acceleration was notoriously slow.

A tank containing 195 gallons of ethyl alcohol rocket fuel was installed above the jet-pipe. Another small tank ahead of the LOX sphere held 11 gallons of a 90 percent solution of hydrogen peroxide to propel a turbo-pump for feeding fuel via two separate centrifugal pumps, one for oxygen and the other for alcohol, to the rocket motor. The turbo-pump, derived from one in the German V2 missile and similar to the one used in the Bell X-2, derived pressure from hydrogen peroxide decomposing over a solid catalyst and releasing energy as steam and oxygen. It exhausted into the main jet-pipe. The solution was so strong and volatile that a cloth soaked in it would spontaneously combust. Further gas containers were needed to pressurize the LOX and alcohol tanks so that the pumps could operate successfully, so nitrogen and helium tanks were fitted behind the LOX tanks for this purpose.

The temperamental LOX, Synasol (denatured alcohol and water mixture), and hydrogen peroxide used by the rocket engine all posed

John Griffith uses two JATO bottles to blast the second D-558-2 aloft. JATO provided a disappointingly meager take-off boost unless four units were used. If one misfired or failed to jettison, its drag and 200lb weight could upset the aircraft's aerodynamic balance. The take-off run was still too long, increasing the dangerous consequences of a tire or undercarriage failure at 190mph. (NASA)

Robert Champine and John Griffith in the second Skyrocket made a series of stability and stalling tests in September and October 1949, and the aircraft is seen here being towed back to the NACA hangar after one of those flights. The turbojet intake flaps are wide open to admit air for the powered landing. With a jet engine only, the Skyrocket's performance was disappointing. Although the Skyrocket weighed twice as much as the D-558-1 and was seven feet longer, its J34-40 jet engine developed only 3,000lb of thrust compared with 5,000lb for the J35-A-11 (TG-180) in the D-558-1. Early flights that required a ground take-off needed a 15-minute car ride for the pilot in the cold, windy pre-sunrise desert conditions to reach the aircraft, which was pre-positioned at the take-off point. (NASA)

severe hazards for those who handled them or flew the rocket-powered aircraft. The skill and care of the maintenance teams prevented many mishaps, but three Bell X-1s and an X-2 with two crew members were lost in accidents in which the instability of the fuel was the main factor. Two North American X-15s were also severely damaged by explosions in their fuel systems. One of them was wrecked on the ground when all of its 16,000lb of propellants exploded due to a frozen valve in one of its fuel tanks. Pilot Scott Crossfield was lucky to escape alive from a 30ft-diameter fireball as the aircraft's cockpit section was shot across the test area at forces of up to 20g with him inside it.

Pilots of the D-558-2 would be more fortunate, as most of the earlier X-1 and X-2 accidents had a single cause, in that the fuel reacted explosively with treated leather gaskets used in the fuel tanks. A safer type of gasket was used for the D-558-2. For the D-558-2 a massive, specialized Rocket Servicing Trailer was provided. It could carry the aircraft back for repair if it was damaged, as it was after its first rocket-propelled flight. It had a shower cubicle so that anyone who became exposed to the highly toxic hydrogen peroxide could be decontaminated and a nitrogen supply so that any fuel substances remaining in the aircraft's fuel system could be neutralized to allow a Skyrocket to be loaded onto the trailer for transportation at up to 40mph.

BUILDING THE BULLET

Douglas used a similar type of construction to that of the Skystreak, but the fuselage was extended by seven feet and the new side-mounted air intake arrangement allowed a more streamlined and bullet-like profile that was more reminiscent of the XS-1. A flush-mounted one-piece cockpit transparency initially fitted to the first Skyrocket reinforced that impression. Overall fuselage diameter was also increased to accommodate the fuel tanks, particularly the spherical container required for the 180 gallons of LOX for the rocket motor.

The fuselage had magnesium skins fastened to aluminum frames and the wing and tail section were made from aluminum. Automatic Handley Page leading-edge slats were installed in the wings to improve low-speed lift, but the flying controls had simple actuating systems without power assistance or trim tabs. Like the Skystreak, D-558-2 had 400 pressure orifices in its wing and tail surfaces. A design team led by Kermit Van Every chose a NACA 63-series airfoil for the wing, tapered from 10 percent thickness/chord ratio at the root to 12 percent at the tip, in view of its known low-speed characteristics. Hydraulic air-brakes were fitted each side of the rear fuselage and the swept-back tailplane used a similar variable incidence mechanism to the Skystreak's. For the D-558-1 it had been necessary to design a main landing gear with very thin wheels to fit into the slender wing, which also contained fuel in its forward sections. This was not possible for the new version, and the landing gear and all fuel tanks had to go in the fuselage. The new aircraft's maximum weight of 16,000lb was almost twice that of the Skystreak, but its maximum load factor was reduced from 18g to 12g at 11,250lb weight. Both aircraft were stressed for Mach 1 at sea level.

Although the Skystreak's emergency escape system had raised many doubts about its practicality, a similar jettisonable nose was used for the D-558-2. Once again, Douglas engineers did not include stabilizing fins, and no suitable braking parachute was available as the system had already added too much weight and complexity. The pilot had to pull a handle that was covered by a glass door to fire the cartridges that separated the nose section, and then wait until the capsule slowed and stopped tumbling before he could jettison the rear bulkhead, exit rearwards, and open his own parachute.

Although this method obviously required several thousand feet of altitude to give the pilot any chance of escape, it was still considered safer than the early "shotgun cartridge" ejection seats. It was the only

The second Skyrocket, fitted with inboard wing fences in October 1949. Four fences were also tested. NACA flights were often infrequent as the aircraft were sometimes grounded for modification or maintenance, particularly towards the end of the program. In September and November 1954 and in March and July 1955 there were no flights, while only one was made in September 1956. (NASA)

D-558-2 BuNo 37973

When it was first rolled out on November 10, 1947, D-558-2 BuNo 37973 had a flush, one-piece cockpit transparency and a short vertical tail. A raised canopy was installed before its first flight and the tail was extended after the tenth flight at Muroc.

practical means of escape once the tight-fitting canopy had been locked down externally, a job for three groundcrew, and it remained an unknown quantity, as no one had to try using it in action. If an emergency occurred within the Skyrocket during an air-launched flight the pilot could blow the canopy off, climb out into the bomber's belly and then ascend a ladder into the bomber's cockpit after the Skyrocket had fallen away. Scott Crossfield was among the test pilots who studied wind-tunnel tests on the escape system, decided that the g-forces experienced in nose separation would be fatal and resolved never to try using the "escape can" system. However, at the time it was widely recommended for all new USAF aircraft.

NACA's standard research instruments included a 12-channel oscillograph to record strain-gauge readings and two 60-cell manometer pressure gauges, all arranged in the D-558-2 mock-up in more useful ways for the technicians than those in the D-558-1 and XS-1. There were more than 900 strain gauges and four miles of wiring. NACA officials had been impressed by the way in which their equipment could take input measurements directly from the aircraft's control system in the D-558-1 rather than from the pilot's controls, recording it all automatically on a 30-channel oscillograph and on film. Selected sections of the flight would be recorded on film at the flick of a data switch in the cockpit, but there was not enough film to cover a whole flight. The cockpit was also similar to the D-558-1's, with a control wheel for the pilot, and it was still cramped. Douglas test pilot Bill Bridgeman thought it was the most confining cockpit he had ever sat in.

The first two examples were completed before the rocket motor was available, so flight trials had to begin with the small J34-WE-40 jet engine only, the rear ends of their fuselages being faired over with conical extensions. When the first was rolled out it still had the flush-mounted cockpit transparency, but before its delivery to Muroc this was replaced with a raised version that was more reminiscent of the D-558-1. Although it was not quite as confining as the canopy fitted to the Skystreak version, its V-shaped cross-section still required the pilot to wear a chamois leather-covered helmet to avoid marking the canopy transparencies. Aircraft No. 2, also without a rocket motor, went to NACA on December 1, 1948 for the beginning of its stability and control trials in May 1949, taking it up to its limiting speed of Mach 0.9 with turbojet power only. The third aircraft remained at the Douglas factory until both types of propulsion could be installed from the outset.

BuNo 37973, covered in tarpaulins that did little to disguise its shape, was taken on a flat-bed truck to Muroc on December 10, 1947 for Douglas chief test pilot John F. "Johnny" Martin, who was most experienced in testing attack bombers and transport types, to make the contractor flights. Bill Bridgeman, who undertook many of the flights in the first aircraft after August 2, 1949, described the reluctance of the firm's engineering pilots to fly the rocket-powered aircraft when invited to do so. It was "never a popular airplane" in his opinion, and one which did not excite him at first glance. When requested to bid for the "hot seat" privilege, they conspired to name unrealistically high amounts for payment. Martin, who was delivering an aircraft to Paris, received a message saying that his bid was awaited. Unaware of the situation back home, he made a sensible bid and was duly selected as the first Skyrocket project pilot.

After the usual taxi tests on January 5 and February 2, 1948, in which low engine rpm was a concern, Martin took the Skyrocket aloft on February 4. The lack of power meant a take-off roll of almost three miles and some lack of stability while the aircraft gained adequate speed. Engine power was obviously severely deficient and actually seemed to reduce during the flight. It provided 30 minutes of powered flight. Apart from deficiencies in the engine itself, it was clear that the narrow air intakes were admitting insufficient airflow and the unusual canted engine exhaust system also reduced engine efficiency. As the design could not be altered to offer remedies in those areas, Douglas resorted to jet-assisted take-off (JATO) solid-

Groundcrew prepare the second aircraft for its October 14, 1949 flight by John F. Griffith with turbojet and rocket power. The turbojet was removed during a long modification period in 1950 and the aircraft was equipped for air-launch, with its first "drop" in January 1951. (NASA)

Although there was much-publicized rivalry between the USAAF/USAF and USN supersonic projects, there was also cooperation, particularly in the early days. The USAAF allowed the Navy access to its plans to air-launch X-planes, while Bell was allowed to use the USN-sponsored LR-8 rocket motor as the basis of the X-1's engine. All four chambers of the rocket unit are alight for this high-speed run. (NASA/Terry Panopalis collection)

fuel rockets on July 13, two of which were installed on the rear fuselage flanks for Martin's flight. Take-off runs with two JATO units were reduced to 8,200ft.

However, once airborne, the climb, dive, and stall tests to date indicated a lack of directional stability, exhibited in a tendency to Dutch roll, particularly on final approach to landing where it could roll 20 degrees left or right repeatedly within seconds. Bob Champine commented that "We got used to it, but it was never very comfortable." That tendency was partly remedied by an 18-inch increase in the height of the vertical stabilizer, which was flight-tested on August 4. JATO was used for most flights while the internal rocket motor was still awaited, and Gene May flew with three JATO "bottles" in November, the third one being mounted in the space allowed for the LR-8 rocket engine's exhaust. Four were trialed in July 1949, adding a total of 4,000lb of thrust for up to 14 seconds per unit to give the aircraft an initial kick towards flight speed. The two additional units were hung between the main landing gear units and they were ignited in pairs.

All were jettisoned when spent, but with jet engine power only test pilot Bill Bridgeman found that the aircraft "handles like a truck, heavy and large." Take-off with a full load of rocket fuel aboard required the brief use of two LR-8 chambers to get airborne, and pilots knew that the jet engine alone would not keep the plane in the air if the rocket motor failed in that situation.

Gene May, veteran of 121 Skystreak flights, took over from John Martin as primary Phase II D-558 pilot on September 16, 1948 and began tests for buffet, stalls, and maneuvering in 37973, call-sign "Navy 973." Meanwhile, the second aircraft was under NACA supervision and its instrumentation package was installed while engineers sorted out ongoing problems with the jet engine.

From May 1949 Bob Champine and John Griffith had shared flight test duties with the number three D-558-1, and Champine was finally able to make the first NACA Skyrocket flight on May

24, 1949. A program of stability and control tests began on June 3. Champine dived to 0.87 Mach, the highest Skyrocket speed attained to date, and by August 3 the only problem was a cockpit recording camera fire which filled the cockpit with smoke. He encountered a more severe problem on his seventh flight on August 3 when the aircraft suddenly pitched up with 5.8g force during a 4.2g turn at Mach 0.6. Alarmed at this sudden demonstration of a severe pitch-up phenomenon that had been only partially forecast during wind-tunnel tests, Champine ended the flight.

He and John Griffith then made 14 more flights to explore the pitch-up characteristics and the aircraft's longitudinal stability. Griffith experienced pitch-up and snap rolls during several high-g turns, and for the aircraft's 19th flight tufts were applied to the tail area, with a camera to record their airflow patterns in conditions approaching a stall. It was also established that at high angles-of-attack the wingtips would stall before the wing roots, and if this happened aft of the aircraft's center of gravity it would cause pitch-up. This part of the program included some risky moments for Griffith as he attempted to push the Skyrocket past the level of unstable flight that May had experienced:

"One of the flights was a series of pull-ups at 200–240kts. In a pull-up, when the airplane got to a pitch-up angle-of-attack it would be interesting to see the position of the horizontal tail in the wing wake. I expect that when the pitch attitude of the airplane was such that the down-wash from the wing went over the horizontal tail it pitched up quite sharply. At 220 to 240kts it wasn't too bad but at maybe 280kts, when I hit that point without me doing anything except pushing against the stick, the airplane pitched up to a stall and a snap roll. It wasn't any problem to pull out of a snap roll, but quite a surprise to be doing a pull-up and all of a sudden the airplane's going out of control."

Uncommanded rolling and yawing occurred on several occasions, and on one low-speed flight at 14,000ft with the landing gear and

The first D-448-2, after conversion to all-rocket power, is given a liquid oxygen leak test at Edwards AFB in 1956. Its elevators and rudder remained unpainted. Although it received full NACA markings, it made only one flight for the Committee (on December 17, 1956 with John McKay) after its prolonged conversion program and over five years after its 122nd Douglas flight. The test program for the first two aircraft was ended in that month and the third Skyrocket finished its tasks on August 28, 1956. (NASA)

ROBERT APGAR CHAMPINE (1921–2003)

Bob Champine was raised in Minneapolis and spent much of his childhood doing menial jobs at the local airport in the hope of being offered a ride in an aircraft. He realized that ambition at the age of 12 and made his first solo flight in a Piper Cub in 1940. After graduation from the University of Minnesota in Aeronautical Engineering in 1943 and subsequent service as a US Navy pilot flying numerous types from the F6F Hellcat to the F9F Panther, Champine joined the NACA Langley Laboratory in December 1947. Among the many aircraft that he tested there under the guidance of Herbert Hoover was the Bell L-39, modified with swept wings to provide data for the Bell X-2, D-558-2 and military types such as the F-86 Sabre.

He soon became used to flying most of the 40 to 50 aircraft types which were typically used at Langley, from B-29 bombers to F-100 Super Sabre fighters. Eventually Champine would have 155 aircraft in his log book and 11,300 flying hours. In many cases Hoover would simply provide him with the pilot's manual the night before a flight in a new aircraft.

Hoover transferred him to the High Speed Flight Research Station at Muroc in October 1948 to fly the Bell XS-1 and both variants of the Douglas D-558 for NACA, replacing Howard Lilly who was killed in a D-558-1 crash. Champine became the sixth pilot (and the third civilian pilot) in the XS-1/D-558 program to reach Mach 1 (in an XS-1) and the first NACA pilot to fly the D-558-2 when he made a 35-minute flight in BuNo 37974 on May 24, 1949. After his return to Langley, where he was instrumental in solving problems with the F8U Crusader's wing-actuating system, Champine became involved in the Mercury, Gemini and Apollo space programs as a NASA test pilot-astronaut. Although his height precluded him from flying in the Mercury capsule, he assisted with the design of high-g couches for space capsule occupants and worked on the Apollo Lunar Excursion Module (LEM) and the space shuttle flight simulator.

In 1979 he retired as the senior research pilot at the Langley Research Center after a career spanning 56 years, with 32 years of test flying. Champine passed away on December 17, 2003, the 100th anniversary of the Wright Brothers' first flight.

flaps extended the aircraft pitched up. Pushed harder, it then entered a spin from which Griffith eventually managed to recover at 7,000ft after raising the flaps and landing gear. "The lake bed was at 2,400ft [altitude]. I think from then on if I was going to do any stalls I would be at 20,000ft."

The solution reflected a problem that was common to many early fast jets where the airflow over the horizontal stabilizers was blanked off by the wing at high angles-of-attack, preventing them from having any real influence over longitudinal control. In their position half-way up the vertical stabilizer, they could be affected in that way at angles-of-attack that would be reached during high-g turns or close-to-stalling speed. These tests ended on January 6, 1950 and the aircraft was returned to Douglas for installation of its LR-8-RM-6 rocket engine and modifications to enable it to be air-launched.

Champine also experienced the aircraft's unpredicted behavior in August 1949 at high speeds and high altitudes when the angle-of-attack was raised.

"If you pulled up and got to 4 or 5g it would suddenly stall in such a manner that the lift distribution on the wing would cause it to pitch-up violently. It would go to extremely high angles-of-attack, between 45 and 60 degrees, and then it would start to roll violently, so the aircraft became completely and totally out of control – just spinning around in the sky. Once you fell into it you had no way of controlling it. You just had to ride it out until you were eventually falling nose-down in a spin. Once you were able to un-stall the wing with nose-down elevator you just used opposite rudder and it would recover in a vertical dive."

Douglas engineers had no real interest in solving the aircraft's handling problems for future production purposes. It was purely a one-off research vehicle, and as long as their tests could be completed, pilots had to live with the plane's eccentricities.

The USN's earlier reservations about air-launching had been largely assuaged in the early part of 1948 by the success of the Bell XS-1 and its B-29 Superfortress carrier. However, no rocket-propelled D-558 flight had yet been achieved, so Douglas and NACA decided to wait until the flight characteristics of the Skyrocket, with its dual power sources, were better understood before seriously considering the air-launching option. The completion of the third Skyrocket, with both jet and rocket engines in place, made that possible. Gene May made its first flight on January 8, 1949 and used its rocket engine for the first time in a flight on February 25. On March 4, he was able to evaluate rocket-powered flight more accurately, albeit in a 25-minute flight that had to be terminated when a fire-warning light illuminated.

After pressurizing the three pressure systems and priming the rocket motor, which had to be completed within ten seconds, the aircraft had blasted off the runway within a mere 2,600ft, lifting off at 157kt with all four rocket chambers burning and the J34 at full power. It climbed to 15,300ft before the rocket fuel was exhausted. Two barrels of the LR-8 were usually lit up when the take-off speed reached

Walter C. Williams (left) discusses a D-558-2 flight with test pilot Scott Crossfield (center) and former test pilot Joseph R. Vensel, who headed NACA's flight operations at Muroc from April 1947. Walt Williams joined NACA in 1940 and became its project engineer on the XS-1 and director of the test program for the D-558-2, among many other ground-breaking projects. (NASA)

100mph. With two alight, the speed rapidly increased, and after take-off at around 190mph, the Skyrocket accelerated to 300mph in a few seconds. At high altitude, in thinner air, the acceleration was even more dramatic. The LR-8 had to be turned off as soon as possible and reversion to the jet engine only (still needed for landing) caused sudden deceleration. The aircraft then struggled to climb at 200mph with most of its fuel still aboard. When the rocket motor was lit at altitude, the jet engine had to be throttled back to avoid over-speeding and flame-out at the increased speed.

When May returned for an emergency landing, not knowing if the fire-warning light indicated a real fire within the aircraft, the right landing gear brake locked on touch-down, blowing the tire and slewing the aircraft off the runway line at a 100-degree angle. On the broad expanse of the Muroc's Rogers Dry Lake area such landings were possible without wrecking an aircraft, and they were far from uncommon.

May flew again on April 27 and, although his flight was again terminated by a glowing fire-warning light, he was able to judge that "ultimate performance of the aircraft will never be achieved with the limited supply of rocket fuel" using ground take-offs. Supersonic flight was not achieved until June 24 in a 32-minute flight that May described as "glassy smooth, placid, quite the smoothest flying I had ever known." However, taking the aircraft to the point where it could begin a supersonic run was a difficult business, and its lack of duration at altitude reduced its potential speed even with both power sources operating, if it had time to reach those parameters. In most cases the rocket fuel ran out moments after the aircraft had reached Mach 1.

A ground-launched flight began before dawn with a slow procession across the desert floor, led by the radio car holding the pilot and project manager and including a line of cars, fire trucks, an ambulance and ahead of them all the Skyrocket on its massive purpose-built support trailer. Usually, take-off would be planned for the exact moment of sunrise to combine cool temperatures with acceptable visibility. A somewhat hazardous take-off procedure was developed whereby the jet engine was used to start the take-off run. Two chambers of the LR-8 were then triggered to provide acceleration, which was just adequate to get airborne and retract the undercarriage. At lighter weights of around 13,700lb, using a jet-only D-558-2, four JATO units could be used, giving a short take-off of about 2,000ft. At 15,300lb four JATO bottles would be fired at 140mph, giving an 8-second boost to 240mph but still requiring a 10,630ft run.

With a full fuel load, jet, LR-8 rocket, and JATO thrust, the aircraft still required up to 15,000ft of ground run to become airborne, placing terrific strain on the fairly fragile undercarriage and putting the pilot at great risk. A tire or wheel-strut failure at that point would certainly have caused a fatal, explosive crash. On balance, NACA calculated that the addition of the rocket motor increased maximum level speed from Mach 0.9 to Mach 0.95 using a ground take-off – a meager gain considering the additional weight, complexity, and cost of the LR-8 system. The case for air-launching had become irresistible.

Before the argument for using a carrier aircraft was finally settled, Gene May attempted to give the USN credit for the first take-off from the ground using rocket power only. However, having won the race to Mach 1, Chuck Yeager heard of this plan and decided to win the rocket take-off accolade for the USAF as well, continuing his intense personal rivalry with May. On January 5, 1949 Yeager and XS-1 program manager Lt Col Jack Ridley gave their orange record-breaker a half-load of rocket fuel. Yeager took off at 200mph with all four chambers of his rocket motor blazing, and reached 23,000ft and Mach 1.03 only 80 seconds later. The extraordinary acceleration blew off the wing flaps and broke the landing gear retracting handle, but Yeager made a safe landing and claimed that the USAF saw him as "a bigger hero for beating the Navy to the punch than for breaking the sound barrier."

DOUGLAS D-558-2 TECHNICAL SPECIFICATIONS			
	A. Turbojet Ground-launched Model	B. Air-launched Rocket-only Model	C. Air-launched Jet and Rocket Model
Power	One Westinghouse J34-40 3,000lb static thrust at 12,500rpm	One RMI XLR-8-RM-6 6,000lb static thrust	One XLR-8-RM-6 and one J34-40 Total: 9,000lb static thrust (both)
Fuel	260 gallons gasoline	345 gallons LOX 378 gallons dilute ethyl alcohol	170 gallons LOX 192 gallons dilute ethyl alcohol 260 gallons gasoline
Weight (lb) - take-off	10,572	Launch: 15,787	Launch: 15,266
Weight (lb) - landing	7,914	9,421	9,500
Max speed	585mph at 20,000ft (Mach 0.825)	1,290mph at 62,000ft (Mach 2)	720mph at 40,000ft (Mach 1.08)
Common dimensions:			

Wingspan: 25ft	Horizontal tail area: 39.9sqft
Wing area: 175sq ft	Elevator area: 9.4sq ft
Flap area: 12.58sq ft	Vertical tail area: 36.6sq ft
Aileron area: 9.8sq ft	Rudder area: 6.15sq ft
Fuselage length: 42ft	Overall height: 12ft 8in.

CHAPTER FOUR

AIR-LAUNCH

In the "passenger" position, the D-558's few inches of ground clearance required pilots of P2B-1S BuNo 84029 *FERTILE MYRTLE* to use a minimum angle-of-attack for take-off, and also for landing if the rocket-ship was still aboard. This August 18, 1953 flight was one of Marion Carl's altitude record attempts. (US Navy via Terry Panopalis)

The advantages of air-launching gradually gained credibility with the USN as the success of the XS-1 program continued. One of its attractions was that it was a cheaper way of achieving sustained high Mach flight than building a new type of aircraft for that purpose. In a June 1949 memo NACA's Research Airplane Projects Leader, Hartley Soulé, reported that the Douglas Company had contacted NACA's Muroc Unit requesting information about the possibility of air-launching the D-558 from a B-29 bomber in order to explore speeds in excess of Mach 1.4. He commented that "it appears desirable to eliminate the jet engine for the test."

Three months later, NACA's Director of Research, Dr Hugh Dryden, recommended that the Skyrocket should be modified for air-launching, using the same basic system as the Bell X-planes. The safety considerations of avoiding a hazardous, fuel-filled ground take-off, together with the research advantages of direct comparison of similar wing sections for the X-1 and D-558-2, were highly persuasive. Pilots were well aware, for example, that a tiny quantity of their hydrogen peroxide propellant could burn a hole in solid concrete.

NACA was also interested in comparing the behavior of the Skyrocket's airfoil with the new bi-convex type that was being developed for the Bell X-2. The D-558 contract was duly amended on November 25 to include conversion of the second and third D-558-2s for air-launching. Aircraft No. 2 was the all-rocket version, delayed by development problems with the LR-8-RM-5 rocket engine, while the third had both rocket and turbojet power. Douglas had already contemplated the benefits of air-launching a

D-558-2 that had extra rocket fuel in place of its turbojet, predicting that it should reach speeds approaching Mach 1.6 at high altitude. For NACA that performance promised a far longer time for instrumented research and for comparison of the swept wing with the straight wing of the X-1 series at similar speeds, weights, and altitudes.

A P2B-1S bomber, the USN's version of the Boeing B-29, was selected for conversion as a carrier aircraft. NACA and aircrew usually referred to it as "the B-29." Four had been acquired by the US Navy as long-range patrol bombers, and BuNo 84029 (formerly 45-21787) was modified by Douglas, nicknamed *FERTILE MYRTLE* by 1953, and given the NACA number 137. Although the B-29 was a capable bearer, the drag and weight of a passenger aircraft below its bomb-bay made climbs to the 32,000ft-plus launch point a slow process that could take up to 90 minutes, with half an hour for the last 3,000–4,000ft alone. It also put a considerable strain on the piston engines, so that power failures were not uncommon. By 1955 *MYRTLE* was the last P2B-1S in the USN and the spare parts and support had run out, requiring a NACA request to the USAF for parts and maintenance totaling $150,000 per year.

D-558-2 No. 2 (BuNo 37974) was to be the all-rocket example, and it was scheduled for modification to air-launch configuration

Substantial areas of the P2B-1S's fuselage had to be removed to accommodate the D-558's tail unit. Long aluminum beams were fitted inside the bomber's fuselage to strengthen those areas. Most of the P2B-1S's six-man crew were mechanics. By 1950 it had a reputation for engine failures and fires, a worry for the crew when carrying a D-558-2 "white bomb" filled with highly explosive fuels. The program was fortunate in experiencing no losses. (US Navy)

By 1954, *FERTILE MYRTLE* (P2B-1S 84029/NACA 137) had an impressive scoreboard of experimental aircraft launches. D-558-2 loading normally used the hydraulic undercarriage lifts that were built into the "mating" area at Edwards AFB. Standard maintenance jacks under the bomber's wings and fuselage could be used, but winds could make this process slightly unstable. It then had to be accomplished inside a building, such as this hangar at the Dryden Flight Research Facility. (NASA)

together with the third Skyrocket. It had its J35 engine, its gasoline tanks and associated fuel and engine control systems removed when the aircraft was returned to Douglas for installation of its LR-8 engine and received much larger fuel tanks, almost doubling the capacity for LOX and ethyl alcohol fuel. The hydrogen peroxide tank was also enlarged. After the two Skyrockets had been modified and tested they were returned to NACA for continued research, arriving in August 1951. They both had new NACA instrumentation including manometers, strain gauges, and pressure orifices installed (what Bill Bridgeman called the aircraft's "nervous system").

For the third Skyrocket, there were fewer changes as it was to retain its dual power system, but it was equipped for air-launching like the second example. Eight-inch mating hooks were installed on the upper fuselage at the forward extreme of the vertical tail fairing and behind the canopy fairing to suspend the Skyrocket partly within the bomb-bay of the P2B-1S. They retracted behind covering doors after launch.

The P2B-1S carrier aircraft's U-1 bomb racks were adapted for the mating hooks. Seven sway braces were fitted to hold its white passenger steadily, and a large area of the bomber's rear fuselage structure was removed to allow the Skyrocket's tail section to fit with its horizontal stabilizer flush with the edges of the larger aircraft's fuselage walls. A similar type of alteration was made to the B-29 and B-50 aircraft that carried the Bell XS-1 and X-2, together with some hefty reinforcement of the fuselage to compensate for the missing sections. Douglas Aircraft carried out both modification processes at El Segundo, including the provision of a gaseous nitrogen flushing system (like the one on the Rocket Servicing Trailer) for the D-558-2's

rocket motor to prevent an explosive build-up of gases while the aircraft was in the captive-carry position.

Loading the Skyrocket below the B-29 required a similar routine to the procedures used for the XS-1 and X-2. The bomber was raised by massive hydraulic jacks beneath each of its undercarriage units and the Skyrocket was towed and manhandled into position below the bomb-bay. At that point the jacks were lowered and the "passenger" was connected to four hoists that were cranked down from the bomb-bay. It was pulled up and latched to the mating hooks on the internal bomb racks. Sway braces were extended to hold the Skyrocket steady and its undercarriage was retracted (leaving an 18-inch clearance between the rocket plane and the runway) while nitrogen and electrical connections were plugged in.

It was understood that if an emergency arose on board the B-29 the research aircraft and its pilot would be dropped once the pilot had illuminated the "ready to drop" light in the bomber's cockpit. Launching research aircraft was often managed from the bomber's left seat, with the co-pilot managing the normal flying tasks.

In a 1952 memorandum for the Boeing Airplane Company NACA described air-launching as:

"Another form of assisted take-off, similar in some ways to the two-stage rocket missile schemes. The principal differences in operation of the D-558-2 [compared with the USAF Bell XS-1] have been that Bill Bridgeman, Douglas test pilot, instead of a Navy pilot, flew the airplane to the maximum speed and altitude, and also that since the airplane was delivered to NACA, the mother [B-29] airplane has also been operated by NACA personnel."

At the drop point, the B-29 had to be kept completely level and steady to avoid damage to the Skyrocket's tail as it left the bomb-bay. Meticulous engineering and preparation minimized the risk of the aircraft getting stuck and dangling from one hook or rising up after launch and colliding with the carrier, but those possibilities always remained.

Still in its US Navy markings, the second Skyrocket lights up its rocket motor. At Muroc the highly efficient D-558-2 support team was run by Al Carder, and the aircraft's hangar also housed 12 mechanics, a project coordinator, eight engineers, two technicians, representatives from Westinghouse and Reaction Motors, and a secretary. Bob Donovan was project engineer. The engineers determined the Skyrocket's flight plans and planned each one a week in advance to fulfill precise research objectives, although pilots could override what they considered dangerous proposals. Many of the flights were focused on establishing the "buffet boundary" speeds either side of the "sound barrier" – the points at which the aircraft would begin to vibrate, particularly when "g" was applied. Flight at Mach 1 in the D-558-2 was usually smooth, but buffet would occur at higher speeds as the boundary layer of air over the wing was disturbed by shock waves. Approaching Mach 1 the left wing would drop, requiring the pilot's full strength to correct it with the ailerons. (US Navy via Terry Panopalis)

D-558-2 COCKPIT

Cockpit configurations varied between individual aircraft and they were frequently modified to meet test requirements or to allow for changes in the aircraft. This is therefore to some extent a generic version.

1. NACA instrument controls
2. Engine master switch
3. Starter switch
4. Rocket fuel controls
5. Fuel shut-off controls
6. Rocket motor controls and jet engine throttle
7. Radio switches
8. Landing gear controls
9. JATO controls
10. Horizontal stabilizer incidence control
11. Air temperature gauge
12. Speed indicator
13. Emergency fuel pump pressure gauge
14. Landing gear and flap position indicator.
15. Rate-of-climb indicator
16. Attitude gyro

17. Fuel remaining indicator
18. Accelerometer
19. Altimeter
20. Directional gyro
21. Tachometer
22. Tailpipe temperature indicator
23. Rocket motor fuel pressure indicator
24. Fire warning panel and test switch
25. Rocket motor pump over-speed warning
26. Helium control pressure
27. Rocket motor oxygen pressure indicator
28. Pilot's breathing oxygen pressure gauge
29. Catalyst pressure gauge (or tailplane angle indicator)
30. Tachometer (early configuration)

31. Fuel pressure gauge
32. Airspeed indicator
33. Mach meter
34. Turn-and-bank indicator
35. Jet engine oil pressure gauge
36. Start control
37. Cabin altimeter
38. Supplementary pressure gauges
39. Voltmeter
40. Cockpit temperature controls
41. Jet engine bearing temperature indicator
42. Slat control
43. Pilot's oxygen supply
44. Cockpit section jettison controls
45. Rudder pedals
46. Control yoke
47. Data control switches
48. Horizontal stabilizer switch

Bill Bridgeman, a USN pilot in World War II and Douglas's Production Test Pilot for aircraft such as the AD Skyraider, replaced Gene May for the air-launched flights after many delays to allow May to make additional flights. After initial misgivings about the aircraft (shared by most of his colleagues), Bridgeman experienced love at first sight on his initial meeting with what he described as the most beautiful airplane that he had ever seen. He made more than 60 flights in the first D-558-2 from August 2, 1949, beginning with a four-JATO bottle take-off and his first rocket flight six weeks later. He concentrated on the unpowered landing phase and on heeding Gene May's warnings about the Dutch roll tendency on approach to landing and the need to make sure the JATO bottles were jettisoned correctly because of their severe drag penalty. Bridgeman's unpowered landing practice would serve him well in October 1950, when fire warning lights terminated three successive flights. He and fellow Douglas test pilot George Jansen flew the newly converted P2B-1S carrier aircraft to Muroc (renamed Edwards AFB in January 1950), and the third Skyrocket arrived by road shortly afterwards.

Gene May had completed his time on the aircraft, including the first supersonic flight by a Skyrocket. Bridgeman's contract was extended and he made the first air-launched flight on September 8, 1950. That excursion was preceded by two weeks of ground tests to make sure that the revised rocket system was reliable and that the Skyrocket would separate cleanly from its bomb-rack release hooks. Captive airborne tests were flown to prove the aircraft's procedure for emergency jettisoning of its rocket fuel. Dyed water was ejected from the hydrogen peroxide tank of BuNo 37975 to study its flow pattern, and it was found some of the liquid was drawn back into the bomb-bay by the slip-stream.

For the first air-launch on September 8, 1950 only the turbojet was used and Bridgeman made a series of air-starts at different altitudes to monitor its temperature. After three more drops (specified by the

Skyrocket No. 2 with its F-86E chase plane, two of which were assigned to NACA's Phase Two Skyrocket tests. F-86E 50-606 (seen here) was loaned by NACA Ames Aeronautical Laboratory. F-86F 52-5426 joined the High Speed Flight Station briefly from June to September 1954 as a D-558-2 chase plane. Both aircraft, Sabre and Skyrocket, used 35-degree wing sweep. Chase pilots stayed close to the D-558 on approach and landing, giving the pilot visual checks on his landing gear and flaps and calling out altitude readings prior to touch-down, as the pilot's restricted visibility made it difficult for him to judge his height above the ground. Usually two chase aircraft were provided for each flight. Throughout a flight the pilot's view through the narrow canopy was essentially confined to the pointed nose ahead of him. Bill Bridgeman used the needle-nose pitot as a stall warning guide. If he saw it vibrating, he knew he had to back off the "g." (NASA)

The cockpit of the second Skyrocket, with its two-handed control column and pilot's oxygen cylinder. Temperature gauges and the lever for the wing slats are in the center and the Mach meter is at the top left. Test pilot Pete Everest found it "comfortable" for his check ride to 65,000ft as he prepared to fly the Bell X-2, although he also stated that it "did not handle too well at high speeds." (US Navy)

USN and NACA) the LR-8 motor in BuNo 37975 was engaged on November 17. At the apex of its climb, Bridgeman pushed the control column forward to level out the aircraft's trajectory at 35,000ft, but the resulting negative g-forces drove the remaining fuel away from its supply lines in the fuel tanks (or "unported" it), causing the system's turbo-pump to over-speed and shut down the motor. During the next flight a similar phenomenon occurred, but this time it was the turbojet that cut out, and it was subsequently recommended that the jet should not be used above 35,000ft.

Bridgeman flew the second Skyrocket with all-rocket power on January 26, 1951. The first seven air-lifts, spread over three months, had been unproductive due to a series of minor problems that were common to many of the Muroc-based rocket-plane flights. Any inconsistencies in the pressurization, temperature, or stability of the volatile substances in the fuel tanks would often result in an aborted flight and dumping the contents, valued at $1,000. With the D-558-2 the problems usually occurred during pressurization of the fuel system when one part of the system would suddenly lose pressure. Often this was caused by frozen moisture in the pipelines. In all cases project director Al Carder had the final say in whether to continue the flight or abort, and he never let his primary concerns over safety override the impatience that the whole team shared through these persistent problems.

Skyrocket BuNo 37974's 24th flight on January 26, 1951 reached the 32,000ft drop altitude and Bridgeman made his final preparations a minute before launch, setting the horizontal stabilizer to 1.5 degrees nose-down, activating the dive brakes, switching on the fan to prevent windshield fogging and performing cockpit checks. George Jansen in the B-29 began the countdown, expecting to pull the release lever moments later. Bridgeman suddenly noticed that pressure in his fuel system was dropping below the safe level. He hastily called an "abort" message and began to shut down the rocket system and prepare the Skyrocket for yet another tethered trip home.

Unfortunately, Jensen was still counting and had kept his microphone open to do so, blanking out Bridgeman's "abort" call. At the very last moment, Bridgeman realized that a drop was imminent. He instantly reverted to the flight plan, revived the systems in about five seconds, primed and lit up the motor after a 3,000ft fall and took the white rocket-ship to 41,000ft before pushing over into a shallow dive and reaching Mach 1.36. As he approached that speed, he found that the control column had no effect on the flight controls and the electric tailplane incidence switch was the only thing that gave him a measure of authority over the hurtling Skyrocket. The flight lasted only 14 minutes, allowing him just enough time to send Jansen a radio message 35 seconds after launch: "God dammit, George – I told you not to drop me!" Sadly, that was not the only problem in that momentous but troubled flight: Bridgeman understandably forgot to turn on the NACA data-recording equipment for most of the flight.

Problems of this nature were rare, although Chuck Yeager survived two consecutive flights where the very similar release process went wrong. A November 3, 1947 XS-1 flight was aborted when one of the shackle pins in the B-29's bomb-racks got stuck, leaving the XS-1 hanging slightly below the bomb-bay. Yeager crawled back into the bomber, and its pilot, Maj Roberto Cardenas, managed to land with the passenger aircraft fractionally clear of the runway. On the following mission the same problem arose and Yeager started to leave the cockpit. As he released his straps the XS-1 suddenly fell clear in a semi-stalled state and he had to bring it under control after a 5,000ft free fall, light its rocket motor, and complete the flight. For future flights, a large red light was installed in the B-29 cockpit for the Skyrocket pilot to light up if he had a last-minute problem.

Bridgeman's flights followed the usual NACA pattern of small, incremental increases in performance together with careful analysis of the results and a specific set of research objectives for each flight. From April 5 to August 7, 1951 he advanced the maximum speed from Mach 1.36 to Mach 1.88 in five successive flights, experiencing increasing lateral and longitudinal instability as he did so. On the April 5 flight the aircraft had rolled left and right to 75 degrees and the ailerons appeared to have no effect, forcing him to exceed his planned push-over altitude. Douglas engineers introduced a rudder lock to assist stability. It worked, and the ailerons could be used fairly effectively again, although the electric tailplane switch remained the most effective tool for supersonic pitch control.

Controls for the rocket motor, radio, flaps, and liquid oxygen priming are located on the left side of BuNo 37975's cockpit. Marion Carl commented on the difficulty of entering and leaving the D-558-2 cockpit when wearing the David Clark full-pressure suit, and found himself "continually annoyed by my inability to grasp and feel things properly." He recommended improved vision through the helmet face-plate and tinted glass to reduce glare. Early air-drop flights were further impeded by a build-up of freezing condensation between the layers of windshield glass. (US Navy)

ONWARDS AND UPWARDS

As the speed boundaries were steadily increased, the Skyrocket was also reaching higher altitudes. For Bridgeman's Mach 1.79 flight to 64,000ft on June 11, the drop altitude was increased to 34,000ft, and he found

FERTILE MYRTLE acquired its nickname and nose-art to acknowledge its habit of frequently dropping offspring. The effort involved is recognized by the beads of perspiration. The aircraft was built as B-29-95-BW 45-21787 in 1945, and it became BuNo 84029 when transferred to the USN in 1947. It was kept airworthy by the American Air Museum in Oakland, California, in the 1970s and 1980s, appearing in a Disney film, *The Last Flight of Noah's Ark*, in 1980. Kermit Weeks purchased the bomber after it was grounded with wing-spar corrosion in 1984, and its nose section was displayed with the white stork and "baby" on a blue background and with a yellow name at his Polk City, Florida, museum, where the rest of the airframe was also stored. (AFFTC)

that his pressure suit's faceplate was almost completely obscured by ice as he dropped from the carrier aircraft. A miniature hand-operated windscreen wiper was devised to remove the ice.

While the performance margins were gradually extended, Douglas officials were aware that NACA was agitating to take over the second D-558-2 now that the manufacturer's trials had proved its airworthiness. However, the company still wanted to realize the aircraft's Mach 2 potential before handing it over to NACA. For the final group of Douglas flights, Bridgeman tried to push the aircraft to its maximum speed. The fuel system was improved by installing a LOX top-off facility in the B-29 to compensate for liquid that boiled off during the ascent. The extra propellant enabled increased maximum achievable altitudes and allowed another Mach increase after the push-over, which itself would be made at only 3g (rather than the usual 7g) to pick up speed more rapidly.

The June 23, 1951 launch was the final USN attempt at Mach 2, but it had to be terminated at Mach 1.85 and 63,000ft when the unstoppable side-to-side Dutch rolling became too severe until Bridgeman stopped it by raising the nose. Instability was reduced in the next flight by loading the aircraft with 0.6g in the push-over maneuver rather than 0.25g as on the previous flight. Bridgeman's helmet face-plate iced up again and the aircraft continued its relentless surge, denying him the ability to make it turn even after shutting down the rocket motor. He tried again on August 7, reaching Mach 1.88 (1,258mph) at 66,000ft, but the fuel ran out before he could make the last push to Mach 2. Bridgeman was convinced that the Skyrocket could still reach Mach 2 "if lateral control can be maintained." He had already beaten the fastest Bell XS-1 record speed by 200mph, but the Mach 2 record was still more than two years away. The difficulty in attaining it almost cost Yeager his life when the same instability threw his Bell X-1A out of control at Mach 2.44 on December 12, 1953 during his final flight in the X-1 series.

Bridgeman's last ride in the D-558-2 on August 15, 1951 did achieve a record, but in high altitude rather than speed, as the USN had allowed one final flight to take the Skyrocket as high as possible. The USN planners' target was 100,000ft, with a speed of only 90mph at the apex of the ascent. Bridgeman pointed out that the Skyrocket stalled at 160mph. He was advised to push over as soon as the rocket fuel ran out to reduce the chances of a stall and spin. The aircraft, repainted in glossy white once again, was dropped as usual and Bridgeman hit all four rocket motor switches and soared aloft,

increasing the angle-of-attack slowly as he passed 43,000ft. The ascent continued into the black edge of space until he reached 79,494ft – an unofficial world record and an opportunity for a unique vision of the planet as he began the descent.

The results of this flight were kept secret for over a year, but by that time the NACA had finally claimed the aircraft and Bridgeman had moved to the imminent Douglas X-3 Stiletto program. His Skyrocket then joined the other two D-558-2s at NACA together with the third D-558-1 and a Bell XS-1. After its 122nd flight on August 14, 1951, the first Skyrocket was sent back to Douglas for conversion to rocket-only power so that it could join the air-launch research program. That process took five years, and after the modification the aircraft flew only once more when John McKay made its proving flight on September 17, 1956. It had made 122 flights in its original jet configuration and it was placed in reserve as the program was already facing cancelation. By 1959 the aircraft was released for use as a recruiting tool, and it eventually found a place in the Planes of Fame Museum collection.

During these flights, a routine was developed that would serve the flight test team well for years. As NACA described it, "a briefing is held at which the flight plan is discussed. Attending are the research project leader, the crew of the mother airplane, the engineer-test pilot of the research airplane, and the radar crew." The latter would track the flight from the ground, recording details of speed, altitude and timing. NACA's emphasis on the term "engineer-test pilot" would perfectly

In August 1953, NACA 144, the second Skyrocket, is towed to its loading position under its "mother ship" ready for one of Marion Carl's record attempts. The P2B-1S is raised on hydraulic hoists beneath its landing gear. Pits for the hoists had to be blasted out of the ground with dynamite. It was possible for the B-29 to land with a $4m D-558-2 aboard in an emergency (if its fuel tanks were safely emptied) rather than jettisoning it. Carl's third flight was terminated when his breathing oxygen ran out at 32,000ft just before launch due to a faulty coupling in the system. The Skyrocket's fuel was jettisoned and the mother aircraft carried the rocket ship safely home. (NASA)

describe the role of Scott Crossfield in the program. Bridgeman too had immersed himself thoroughly in aeronautical engineering theory before flying the Skyrocket, with particular emphasis on a textbook titled *Stability and Control*. Also present at a briefing was the pilot of the USAF chase aircraft, "whose mission is to observe the test and report any malfunctions, damage or other unusual conditions not visible to the pilot of the test airplane."

For a typical flight, NACA stated:

"The take-off is made as soon as possible after fueling to avoid excessive boiling off of the liquid oxygen. The test pilot rides in the mother airplane during take-off and initial climb. Usually he enters the test airplane at about 10,000ft and begins preparations for the drop. Continuous radio contact is used between the mother airplane, the test airplane, the escort [chase] airplane, the radar tracking station and the ground control station, where personnel are stationed to assist the pilot and to suggest last-minute alterations in the flight plan because of changing weather conditions, airplane functioning, etc."

NEW CREW

Scott Crossfield became NACA's pilot for the D-558-2, as Bob Champine and John Griffith had taken on new roles elsewhere. He was joined by Walter Jones, who began D-558-2 flights in August 1951. Crossfield's familiarization flight in the third aircraft took place on December 22, 1950, and it proved to be an exciting one. Blasting away from *FERTILE MYRTLE* on full jet and rocket power, he climbed to 43,000ft, where the jet engine flamed out, beaten by the low pressure at that altitude. This cut out all electrical power, and shortly afterwards the LR-8 motor stopped when its fuel was unported in a negative g pitch-down motion. When the aircraft battery also failed to supply power due to a faulty relay, Crossfield was left with an iced-up windshield for his dead-stick glide home. "The only choice I had was to put the sun on a spot in the windshield and then fly the airplane so it stayed there and so I knew at least the airplane was the right side up." Fortunately, his chase pilots John Conrad and "Fitz" Fulton were in position to guide him to a safe landing.

After NACA officially took delivery of Skyrockets BuNos 37973 and 37974 on August 31, a two-week task of installing strain gauges, manometers, and pressure orifices for its research schedules began. From March 1951 to September 1955, Crossfield and Jones worked through NACA's systematic program of tests using the second and third Skyrockets, although Crossfield undertook the majority of the flights, particularly those in BuNo 37974. In December 1951, NACA Aeronautical Engineer Herman O. Ankenbruck reported that data obtained from the preliminary flights in this aircraft showed that its "dynamic lateral stability is very poor in the transonic and supersonic speed range and at speeds above Mach 1.4. Static directional stability deteriorates rapidly as the Mach number increases." He stated that NACA was exploring ways of dealing with the problem, of which "the

most promising appears to be the installation of a rate autopilot to add damping in yaw." He also noted that work on a rudder-locking device was continuing to cure rudder vibration at high speeds.

Crossfield and Jones continued to grapple with these crucial problems of stability and control and the pitch-up problem, which still plagued many attempts to push the speed margins yet higher. Buffeting at near-sonic speed normally increased markedly in a turn. A flight by Walter Jones on September 18, 1951 in BuNo 37975, which began its NACA flights on December 22, 1950, ended in a spin after the rocket motor failed to light and he had to jettison the rocket fuel. Continuing with the jet engine only, he attempted an accelerated pitching motion with the landing gear and flaps extended, and the aircraft fell out of control at a high angle-of-attack until Jones cleaned it up and initiated spin recovery.

NACA focused on modifications to the wing of BuNo 37975, adding a nine-inch chord extension to the four feet of each wing in February 1953 in place of leading-edge slats. The aim was to alleviate stalling at the wingtips, which were situated aft of the aircraft's center of gravity and stalled before the wing root, thereby inducing pitch-up. Two flights in February and April 1953 showed that the extensions actually made the problem worse, particularly in turns at high subsonic speeds and 1g stalls. Slightly better results came in mid-1952 from flying with the wing slats locked open or half-open, and Crossfield experienced a 36-degree pitch-up angle at Mach 0.97 on his October 8, 1952 flight with the half-open position. In December NACA reported that "the configurations flown with slats fully extended were an improvement over the original airplane configuration in as much that the pitch-up previously encountered appeared to be eliminated or alleviated except at Mach 0.8 to 0.85."

Another attempted solution involved the installation of wing fences. At first they were placed at 36 percent of the span of each wing and then a second pair was added at the 73 percent span point. The

D-558-2 BuNo 37374 at Edwards AFB after conversion to all-rocket power and air-launch configuration. Its return to NACA as 144 coincided with the painting of a yellow tail-band in place of the previous small NACA shield on the tail. (NASA)

inboard fences were tested with and without slats, and in flights from September 26, 1951 the outboard set did slightly alleviate pitch-up at speeds approaching Mach 0.95. However, they were removed in November 1951. The root of the problem was still the placement of the horizontal stabilizer halfway up the vertical fin. Although it was a typical configuration at the time and a good way of keeping the elevators free of wing-wash in level flight, it was found that at high angles-of-attack the airflow over the elevators was disturbed. In a pitch-up condition, this effect could be sufficient to blank out the horizontal stabilizer altogether. For many subsequent fighter designs such as the F-100 Super Sabre, English Electric Lightning, and F-105 Thunderchief, the solution was to place the stabilizer near the base of the rear fuselage, but cost and structural factors ruled out this type of redesign for the D-558-2.

In another program, the LR-8 motor was tweaked to provide more thrust. In August 1952 Walt Williams asked for up to 50 percent extra thrust to overcome the drag that would result from testing external stores. They would otherwise limit the aircraft's speed to Mach 1 at best. He also hoped that increased thrust would greatly increase the aircraft's performance at the very high altitudes it could reach with a pilot wearing a full-pressure suit. Most of the rocket motor's working parts were steadily improved to give greater reliability, but the most obvious benefit came from extending the exhaust nozzle to give more power at high altitudes.

It was found that the rudder vibration that had affected Bridgeman's flights in the summer of 1951 was caused by a shock wave created around the rocket efflux, and the NACA-built, uncooled nozzle extensions cured this problem and saved complete redesign of the cylinder nozzles. NACA reported that they also added an extra 500lb thrust which, together with improved efficiency from internal redesign, gave the revised LR-8-RM-8 motor a total of 2,000lb additional thrust at 70,000ft altitude, or 7,550lb thrust at 50,000ft. The improved parts could be installed in the engine during routine maintenance as they were generally interchangeable with the existing components. In terms of flight time, the modifications conferred another 17 seconds of powered flight, but that was long enough to boost the speed significantly. It also reignited the USN's desire to reach Mach 2 before Chuck Yeager in the new Bell X-1A could snatch that record.

CARL RETURNS

Col Marion Carl, world speed record breaker in the D-558-1 in August 1947, was recalled to Edwards AFB in July 1953 to make high-altitude test flights of the USN's full-pressure suit. It was hoped unofficially that this assignment might also lead to new record altitude or speed achievements. Carl made two familiarization flights in the third Skyrocket in July 1953, but the third flight in BuNo 37975 was aborted. Carl reported, "At about 32,000ft and a couple of minutes prior to launch the pilot ran out of oxygen. A quick check revealed zero pressure on the D-558 oxygen system. As the cabin pressure was about 15,000ft, the

[helmet] face piece was opened and normal breathing resumed. All lox and fuel was jettisoned and the B-29 landed with the D-558 aboard." For future flights, the pilot had the additional encumbrance of an emergency oxygen bottle strapped to his leg.

Carl then made a series of all-rocket flights to high altitude in BuNo 37974, for which he had to make the first flights in the tightly fitting cockpit wearing the clumsy David Clark Company full-pressure suit that NACA required for flights above 45,000ft, where survival would otherwise be impossible. At the much greater heights that zoom-climbing Skyrocket pilots could reach, the temperature fell below minus 50°F and the atmospheric pressure was around 2 percent of the normal sea level pressure. In the launch aircraft, the pilot required two assistants to help him enter or exit the suit. A knife was provided so that he could cut his way out in an extreme emergency. Carl recalled that:

"After the suit was properly fitted [tailored to each individual pilot] two-and-a-half days were spent in indoctrination on the wearing and functioning of the suit. As the suit fitted almost skin-tight, the first problem was fighting off claustrophobia, which turned out to be quite an item when the pilot became too warm. One of the biggest problems was the control of body heat. The pilot was either too hot or too cold."

Wearing the suit, which inflated at around 42,000ft, also required a different "pressure breathing" technique in which the pilot had to

In NACA's high-speed research stable on August 4, 1953 the third D-558-1 (center left) and a D-558-2 (center right) share the shot with the Convair XF-92A (top left) and swing-wing Bell X-5 (top right). In the foreground are the Bell X-1A (left) and Northrop X-4 Bantam tailless, twin-jet research aircraft, whose handling characteristics prompted Walt Williams to call it a "lemon." An even more challenging type, the Douglas X-3, occupies the center ground. (NASA)

breathe out, driving the oxygen from his lungs after it had been inhaled forcibly as a result of the pressure in the helmet.

For his flights, Carl noted that in the B-29, "immediately after take-off the pilot shed his clothes and donned special underwear." With assistance, he then:

"Climbed into the pressure suit. At about 7,000ft the pilot was ready and started entry into the D-558-2. At about 10,000ft the connections were all made, the canopy was closed and the 'ready to launch' light was turned on. From this point until about five minutes prior to launch the pilot was merely along for the ride and did almost nothing. At five minutes prior to launch the pilot started his check-off [list] with Mr Crossfield on the intercom and performed all steps as required up to the drop. At the drop the only remaining item was [priming and] lighting off the rocket chambers. The No. 3 chamber of the rocket motor was always touched off first, then Nos. 4, 1 and 2. The first two were touched off as fast as chamber pressure permitted or within ten seconds after launch."

After two chambers fired, the aircraft would start to climb, and Carl estimated that "the hardest part of each flight was that of maintaining the proper attitude in the climb. There was no attitude gyro or artificial horizon in the cockpit and the pilot could see neither the ground nor horizon. Movement within the cockpit was extremely limited." Carl's 6ft 2in. height brought the suit's helmet into uncomfortable contact with the small cockpit canopy throughout his time in the cockpit. "It was possible for the pilot to turn his head about 30 degrees each way. [The suit restricted this to 45 degrees in any case]. The top of his helmet touched the canopy and his shins were up against the instrument panel so that only the tips of the toes touched the pedals." This cramped position obscured his view of most of the engine instruments. The control column could not be pulled back to its fullest extent, precluding the use of full-up elevator. In Carl's opinion, if the pressure suit inflated due to loss of cockpit pressure at altitude the pilot would be able to maintain control only by using the electrically powered stabilizer until the aircraft descended to 35,000ft, where the suit could deflate.

Two unsuccessful attempts were made at the altitude record on August 14 and 18, 1953, but on 21 August he was launched at 32,000ft later in the day when conditions were cooler. Following a longer fall after launch because the first rocket chamber would not immediately light, the aircraft climbed smoothly without buffet and the rocket fuel ran out at 75,000ft. The aircraft zoomed to 83,235ft, a new record, and Carl reached Mach 1.5 after the push-over (at only 250mph) as he descended.

The speed record proved to be more elusive. On August 31, Carl's first attempt was aborted when the second Skyrocket (NACA 144) became severely unstable at Mach 1.5, beginning "violent oscillations which the pilot was unable to control." Another attempt on September 2 advanced the speed to Mach 1.728, but Carl dived too steeply, losing too much altitude to achieve optimum speed. He was not asked to try again. Partly, this was a reflection of NACA's concentration on its pure

At several points between 1951 and 1953, various wing slats, fences, and chord extensions were tested with NACA 144 to deal with the Skyrocket's tendency to pitch-up prior to stalling. Scott Crossfield and Walter Jones flew the 1953 tests, including this chord extension. The modifications offered little or no improvement. Just ahead of the vertical fin, one of the retractable mating hooks for suspension beneath the P2B-1S is protruding through its small covering doors. (NASA)

research, which could be conducted satisfactorily at altitudes below 80,000ft and did not benefit directly from hitting Mach 2.

Hugh Dryden at NACA was content to let the USAF reach that speed if they wished. However, Crossfield and Walter Williams persuaded Dr Dryden that there would be scientific benefits in pushing the D-558-2, with its improved engine, to the limits of its performance. Crossfield later claimed that he "just dropped the hint to the Navy that wouldn't it be great if they could whip Yeager's ass and beat him to Mach 2!" With further persuasion from Rear Admiral Apollo Soucek of the Bureau of Aeronautics and Marion Carl, the Pentagon agreed to a one-off Navy attempt at Mach 2 before surrendering the task to Yeager.

DOUBLE-MACH D-558-2

Crossfield began test flights on September 17, 1953, with the nozzle extensions installed and a tank regulator in the cockpit so that he could increase pressures in the four rocket chambers. The two modifications increased total thrust to almost 9,000lb. Flights continued through October and November as stability and aerodynamic load tests, but with gradually increasing maximum speeds. In mid-November NACA plotted a flight profile in which Crossfield would have to follow a very precise climb, speed, and angle-of-attack after launch. After the "push-over" he would enter a shallow dive and reach Mach 2.

NACA 144 (BuNo 37974) was carefully prepared with waxed finish, sealed panels, and a prolonged "cold soak" after the LOX was loaded to allow the liquid gas to settle and be topped up, both on the ground and from the carrier aircraft's top-up system when airborne. The process also added seven seconds to the burn time. At the tail end, the stainless-steel propellant vent tubes were replaced with lighter aluminum versions, bent so that they would burn off in the rocket efflux and save a little weight.

On November 20, 1953 the mighty P2B-1S and its load took off with NACA pilot Stanley Butchart at the bomber's controls. Butchart was a former World War II Naval Aviator in a torpedo-bomber squadron with future US president George Bush. He spent 25 years in research aviation, flying the D-558-1 on 12 occasions, and he was the High-Speed Flight Station's main multi-engined pilot, launching hundreds of experimental aircraft from B-29s and other carrier aircraft for six years.

At 10,000ft altitude Crossfield prepared to launch, and on release he ignited all four rocket chambers and climbed to 72,178ft before pushing over into level flight in the plan plotted by Herman Ankenbruck and accelerating for another 45 seconds until the rocket fuel was exhausted. At that point he was surprised to notice Mach 2.04 on his Mach meter. He had become the world's fastest man and the D-558-2 had at last reached Mach 2, far beyond its maximum design speed, and the USN held the records for both speed and altitude.

Three weeks later, Yeager boarded the modified Bell X-1A, determined to beat Crossfield's record substantially in a contractor flight effort unofficially called "Operation NACA Weep". Yeager's team estimated that the X-1A's horizontal tail would be effective up to Mach 2.3 at 72,000ft, but its instability problems had already been seen at Mach 1.8. On December 12, Yeager launched at 30,500ft, accelerated to Mach 2 at 76,000ft, pushed over and entered a shallow dive that reached Mach 2.4 before all the flying controls lost effect. The aircraft tumbled, completely out of control in a wild spin for 50,000ft, battering him into near unconsciousness before he could recover it. Fortunately, the X-1A's immensely tough airframe remained intact and Yeager also survived, commenting, "Boy, I'm not gonna do that again." It was his last rocket-plane flight.

Crossfield continued the stability research flights until October 1955 using both NACA aircraft, but during that period three other well-known pilots, Lt Col Frank "Pete" Everest, Joe Walker, and Capt John McKay checked out in the Skyrocket. McKay made the majority of the remaining D-558-2 flights, while Everest's two flights were preparation for his part in the Bell X-2 program.

McKay's flight in NACA 144 on March 22, 1956 was particularly eventful. Stan Butchart and future astronaut Neil Armstrong were flying the P2B-1S and they had struggled up to 31,000ft, ready to launch McKay in the rocket plane, when the bomber's No. 4 engine gave up. After hurried conversations with flight engineer Joseph Tipton, the prop-feathering button was pushed as the failed engine's propeller seemed to be stationary. Instead, the propeller began to turn again with increasing speed despite two more attempts at feathering it. The hydraulic tank for feathering propellers only allowed three attempts. At the same time McKay reported that a nitrogen valve in the cockpit supplying his rocket motor had broken, effectively aborting the mission. However, in accordance with the established safety drill, the Skyrocket had to be dropped in an emergency. The P2B-1S was close to the point where the drop was scheduled, and Butchart decided to make the turn and cross the drop-point as planned. It was also necessary to dive the aircraft so that it would exceed the minimum speed of 210 to 220mph below which the Skyrocket would stall when dropped.

WILLIAM BARTON BRIDGEMAN 1916–78

Born in Ottumwa, Ohio, to a barnstorming flying circus pilot and his young wife, who separated soon after his birth, Bridgeman was brought up in Malibu, California, by his grandmother. She instilled in him a strong sense of independence and self-sufficiency and enabled him to enter university to study geology. He enlisted in the US Navy and, as a member of a USN PBY Catalina squadron at Pearl Harbor in December 1941 when the Japanese navy struck the base, he flew the last remaining undamaged PBY flying-boat in search of the enemy fleet. He then flew PB4Y-1 Liberators with VPB-109 "Reluctant Raiders" in combat from various central Pacific island bases until its disestablishment in October 1945.

Postwar, Bridgeman flew DC-3s for Trans-Pacific Airlines. Seeking a more exciting career, he joined Douglas in 1949 as a production test pilot for AD Skyraiders. The company was struggling to find enough test pilots for its other new products, the A2D Skyshark, F4D Skyray, and F3D Skyknight, as well as the X-3 Stiletto and the D-558. He was offered the post of D-558-2 project pilot, replacing Gene May. He quickly became jet-qualified on a P-80 fighter and spent countless hours with flight manuals. Bridgeman was told that he was "biting off a real mouthful" in accepting the job, but once again the sense of adventure and freedom appealed to him despite his lack of relevant experience in high-speed flight. During his time with the Skyrocket he set two world speed records in 1951 and broke the world altitude record. He remained with the program until August 1951, averaging one Skyrocket flight per week. He also flew the Douglas X-3's contractor flight program and applied for a USAF astronaut program, although that was canceled in 1958. After several years as a commercial test pilot for Grumman Aircraft, he went into the real estate business but also flew a Grumman Goose air-taxi service from

Bill Bridgeman (Author's collection)

Los Angeles. He was lost at sea in the crash of that aircraft in September 1968.

While Armstrong made the shallow dive, Butchart attempted to pull the emergency release T-handle to release the Skyrocket, but nothing happened. He resorted to the two "pickle switches," normally used to drop bombs but adapted as back-up release for the rocket-plane cargo, which then fell away cleanly and made a safe landing.

Moments after the Skyrocket plunged away, the bomber's engine "blew – big-time," according to Butchart. The propeller nose dome flew ahead of them, "looking like the kitchen sink going by, and the blades went in all four directions." One sliced through the No. 3 engine nacelle and continued on through the bomb-bay area where the Skyrocket's cockpit had been suspended seconds before, hitting the No. 2 engine on the P2B-1S's other wing. Fragments severed aileron cables together with the fuel lines and throttle cable to the

The battered P2B-1S after its near-disaster on March 22, 1956. When the propeller from the No. 4 (outboard) engine flew off and disintegrated, it cut into the No. 3 nacelle and wrecked that engine too, forcing pilot Stan Butchart to feather it. Fragments penetrated the fuselage and one propeller blade passed clean through the fuselage where John McKay had been sitting in D-4558-2 BuNo 37974 moments before, ready to make the aircraft's 67th NACA flight. Repairs took five months. (NASA)

No. 3 engine, which had to be feathered. The two pilots then had to descend from 30,000ft and aim for their one possible attempt at landing on the lake-bed, using only the left inboard engine. The left outboard engine (No. 1, which had to be replaced during subsequent repairs) created too much torque and had to be throttled back. With such asymmetric power, it took both pilots to operate the elevators and rudder, with Armstrong working the ailerons, to get the battered B-29 down safely after the most potentially disastrous Skyrocket launch.

McKay made eight more flights in NACA 144 after the P2B-1S had been repaired, completing tests on vertical tail loads, dynamic stability, and sound pressure levels in the aft fuselage until the aircraft was grounded on December 20, 1956 and its NACA instrumentation was removed. He also flew NACA 145 for the final stages of its program, making 14 flights, which included tests for wing-load data with external stores on two pylons under the aircraft's wings.

Ed Heinemann was a pioneer in designing low-drag tanks and pods for use beneath fast aircraft, and this research extended into the design of low-drag Mk 80-series bombs, using similar shapes. The initial tests used shapes that were 15ft long and held 150 gallons of fuel (although it also resembled a Mk 7 nuclear weapon), while a second version only 10ft long simulated a 1,000lb bomb, complete with four tail-fins. The pylons and wiring were installed on NACA 145 and Crossfield made the first 30-minute test flight on June 2, 1954 with the "Mk 80" type, having first flown with the pylons only. He reached Mach 0.72 and noticed only slight reductions in handling and performance.

CHAPTER FIVE

SUPER SKYROCKETS

Although the third stage of the Douglas jet-rocket program was to have produced a full-scale operational jet fighter, the USN quietly dropped that requirement. Instead, the very different Model 671 (also unofficially referred to as the D-558-3) originated in 1954 when the Office of Naval Research asked Douglas for ideas for a far more ambitious research aircraft than the D-558-2. Using rocket power, it was intended to be launched from a B-52 carrier and attain hypersonic speeds (above Mach 5) and an unprecedented altitude of one million feet. NACA had air-launched a T40 solid-fuel rocket in March 1953 that exceeded Mach 5 for the first time, and a manned aircraft seemed to be a feasible next stage, using a 50,000lb-thrust XLR-30 rocket motor. Although the Model 671 only reached the predesign stage, it evolved from the D-558-1's straight wing combined with a modified version of the D-558-2 fuselage.

Douglas received little more specific data on the USN's requirements, but its design team focused on the very demanding altitude requirement rather than speed. There were immediate reservations about the feasibility of re-entering denser atmosphere from such a high altitude without losing control or suffering destruction through extreme overheating at 3,300°F. By the end of 1954 it became clear that the project was to be superseded by a joint service hypersonic research aircraft venture aimed at Mach 6.7 speed and a more feasible altitude of 250,000ft.

In some ways the natural successor to the Skyrocket would have been the Douglas MX-656, which became the Model 499D or X-3 in 1949. From May 1945, in parallel with the D-558-1, Douglas worked on its

Douglas-built 150-gallon drop tanks were tested on the third Skyrocket in a stores program that began in February 1954. The fuel tanks, which resembled Mk7 tactical nuclear weapons, were flown in October 1954 and May 1955, and the 1,000lb store that introduced the shape of the Mk 80 series of bombs began flight tests in June 1954. (NASA)

A D-558-2 pilot would light the rocket motor as soon as the aircraft had cleared its carrier position. The LR-8-RM-6 motor had to be removed after 30 minutes burn time and returned to Reaction Motors Inc. for a complete overhaul. The company delivered enough parts to assemble 21 engines, so that replacements were always available. The most common cause of damage was a failure in the cooling system for the rocket tubes, in which temperatures could reach 6,000°F. When the motor was shut down after a static test, water had to be squirted into the four barrels to extinguish the residual fires. The pilot's job involved close attention to his row of pressure dials for the rocket motor during a ground-test, despite the painfully high noise levels, reduced to a low rumble within the cockpit in flight. Pressure had to be kept within the range of 600–2,000lb. (US Navy via Terry Panopalis)

aptly named supersonic X-3 Stiletto. Two examples were ordered in June 1945, but only one (48-2892) was completed. Its development program was slow and troubled by lack of engine power and relevant research data, and in 1950 Walt Williams, who managed NACA's research flight programs from the earliest days at Muroc to the time of the space shuttle, warned prophetically that it was "very complicated and it might turn out to be a dud." Although its futuristic appearance made it look as if it could reach the sustained ten minutes at Mach 2 goal that its designers had in mind more easily than a D-558-2, its J34 interim engines developed less than half the power that it needed and its handling characteristics elicited the comment from test pilot Pete Everest that it was "the most difficult airplane I have ever flown." He flew 122 different types.

In Scott Crossfield's opinion, "no more complicated, botched-up, dangerous airplane was ever produced." Bill Bridgeman was appointed as its chief test pilot for the single completed example and made its very belated first flight in October 1952, followed by 20 more flights. His first impression of the X-3 mock-up with twin 3,000lb-thrust J34 engines in a 66ft 9in. needle-nosed fuselage and a 22ft wingspan of only 166sq ft in area echoed the words of a contemporary Douglas advertisement. He witnessed an aircraft "as long as a DC-3 with a wingspan similar to a DC-3's tailplane," but to him it was an "unholy-looking machine."

Like its other pilots, Bridgeman found that merely persuading it into the air was a challenge due to inadequate thrust. In its early stages rocket and wingtip-mounted ramjet power had been considered, as was water injection or hotter afterburning for its two 3,370lb dry-thrust Westinghouse 24C (J34-WE-17) engines to compensate for delays in developing its intended J46 engine. Westinghouse's 24C-10 (J46) engine, which was expected to propel the Stiletto to Mach 2 at 35,000ft with its 6,600lb of afterburning thrust, was not expected until 1955. J34s were similar to the model originally used in the D-558-2, and they produced grossly inadequate power for the X-3's intended flight program. Crucially, the J46 never materialized, and would in any case have been too large for the X-3 airframe. Ramjet or twin XLR-11 rocket installations were not pursued on the assumption that the more powerful turbojet could be used. The absence of the J46 also delayed construction of the second aircraft, which was eventually used only for spare parts.

After rejecting a separating nose capsule like the D-558's for the pilot to escape, Douglas opted for a downward ejection seat. The pilot was pulled up into the cockpit, from which visibility was very poor, by an electric elevator system. The Stiletto was designed to sustain Mach 2 speeds for at least 30 minutes, but its minimal wingspan, intended to minimize supersonic drag, made landings at a scorching 250mph inevitable and reduced maneuverability considerably.

Bill Bridgeman made its 26 contractor flights, reaching a maximum speed of Mach 1.21 in a 30-degree dive and managing a single-engine landing on the 23rd flight by using afterburner to keep the X-3 airborne for an approach to landing. Transferred to NACA, who intended the aircraft to carry a large 1,200lb instrument package, it was assigned to pilot Joe Walker. He encountered the dangerous roll divergence or inertia coupling ("totally out of control") phenomenon, and analysis of those flights did help in designing supersonic aircraft with more stable flight characteristics. However, by 1952 the X-3

Scott Crossfield gives Walt Williams the first report on his Mach 2 flight on November 20, 1953. During the test program, he attempted to shorten the Skyrocket's landing run sufficiently to use Edwards AFB's paved runway during the winter when the dry lake was temporarily covered in rain water. He designed a small brake chute for installation in the third aircraft, but inadequate electrical power was available to deploy it properly for its sole test flight and NACA dropped the idea. (NASA)

After withdrawal from NACA service, the three Skyrockets were taped up for storage on the NACA ramp at Edwards AFB in 1957.

was superseded by more useful research tools, including much more practical production military types. It contributed to the understanding of roll-coupling instability problems to some extent, and it provided early experience in the use of titanium in airframe structures. Using a one-piece horizontal stabilizer yielded useful data for future fighter designs. Its tiny wing and downward ejection system also helped Lockheed in designing the F-104 Starfighter, which immediately became a successful supersonic design as its intended and very powerful J79 engine was available early in its development. The X-3 also pioneered the use of the ribbon-type braking parachute, but its take-off speed, the highest for any aircraft at the time at around 240mph, made it a hazardous and extremely unpopular aircraft to fly.

HEINEMANN'S HYPERSONICS

In May 1955 Douglas, in competition with North American, Bell, and Republic, entered a new design for hypersonic research. The original Model 671 from 1954, which was essentially similar to the D-558-2, was enlarged, a jettisonable ventral fin was added and the horizontal stabilizer was moved to the rear fuselage. It thus became the Model 684, but the often-applied designation "Skyflash" name was never officially used for it.

The aircraft had a lightweight titanium and magnesium airframe built around a 50,000lb-thrust Reaction Motors XLR-30 Super Viking rocket motor and its fuel tanks. An ejectable nose section was retained

DOUGLAS MODEL 684 "SKYFLASH"

The Douglas Model 684 was a projected development of
the D-558-2 for hypersonic research. With a length of
46.7ft and a 19.5ft wingspan, it could have been carried
by a B-29 or B-50 like earlier X-planes.

and the pilot would have parachuted to the
ground inside it. The airframe was to be
coated with an ablative material that would
have protected the surfaces at hypersonic
speed, burning off gradually. Overall
temperatures of 1,400°F were anticipated
on re-entry to the denser atmosphere,
and the heavy, 0.75in.-thick magnesium
skin was meant to act as a heat sink. The
copper leading edges of the magnesium/
thorium-zirconium flying surfaces,
where temperatures of over 3,300°F were
anticipated, were to be protected by chilled
liquid oxygen or water spray to absorb
the worst of the heat. Chief Designer Ed
Heinemann anticipated that they would
"actually glow in the dark, but prevent
the magnesium from reaching a critical
temperature limit by the time the dive was
completed."

An innovative control system, using
hydrogen peroxide-powered thruster
jets in the wingtips and tail, replaced the
conventional control surfaces at high
altitudes where the thin atmosphere would
not allow them to operate normally. Brief
bursts of vapor could be used to correct the
aircraft's direction and angle-of-attack. This
was particularly necessary for the re-entry
dive from the ionosphere to avoid the
aircraft entering the atmosphere at anything

In the Douglas X-3's design stage, air-launching and rocket boosting were considered. A jettisonable nose section was suggested, with additional canard surfaces for stabilization, for the pilot to escape in an emergency. The chosen downward-ejection system with retractable stabilizing fins, which was thought to put less strain on the body than a contemporary upward ejection seat, precluded an opening cockpit canopy. Performance estimates included sustained Mach 2 flight at 35,000ft, but only if the Westinghouse J46 engines could be used. Air-launching from a B-36 Peacemaker was considered but rejected as too costly. The tiny wing was too thin to accommodate aileron and flap actuators, which had to be placed in external fairings. (AFFTC)

other than a nose-first angle. Auto-stabilization via an autopilot was included to control the aircraft at high supersonic speeds and the flying controls were power-boosted. Unlike the other competitors, Douglas included a conventional tricycle undercarriage rather than the skids used as main gear on the North American Aviation proposal.

Douglas predicted that the Model 684 would be light enough to travel aloft beneath a B-50 carrier aircraft, whereas the larger, heavier North American design (which became the X-15) would require a converted Boeing B-52 bomber with a specially adapted underwing pylon mounting. A D-558-3 would have spanned only 19.5ft, with a fuselage 46.6ft long and a distance of 13ft to the top of its tail fin. Its maximum take-off weight was 22,000lb (about twice the weight of a loaded D-558-2), of which 15,000lb was liquid oxygen and ammonia fuel.

LEFT
On April 10, 1958 Douglas issued this artist's impression of the D-558-3 "high altitude design study." (Terry Panopalis collection)

ALBERT SCOTT CROSSFIELD (1921–2006)

Born in Berkeley, California, Crossfield was a US Navy F6F Hellcat and F4U Corsair pilot during World War II. He studied aeronautical engineering at the University of Washington with Stan Butchart, specializing in wind-tunnel technology. With a Master of Science degree secured, he joined NACA in 1950 as a research pilot, making 99 flights in the X-1 and D-558 and many others in aircraft like the F-100 and F-86, in which he studied the roll-coupling phenomenon. Crossfield made up to three flights daily in aircraft such as the X-4 and X-5, but his "biggest effort was reserved for the Douglas Skyrocket." Working with NACA at Edwards AFB, he had a key role in the development of the first full-pressure suit, which led to a new generation of astronauts' suits for the US space program. For the prototype, "much of it was sewn on my wife's sewing machine." With more experience of rocket planes than any other pilot, Crossfield left NACA in 1955 to work on design and introduction of the X-15 with North American Aviation as chief engineering test pilot. He made the first 14 proving flights in the X-15, reaching Mach 2.97 and an altitude of 88,116ft.

In five years of test-flying, he achieved the highest total of rocket-plane flights of any pilot and over 220 flights in all, totaling 600 hours. Crossfield then moved on to a new generation of North American projects including the Hound Dog missile, Saturn rocket booster, and the Apollo command and service modules. Crossfield then moved in 1967 to an executive role at Eastern Airlines and in 1974 to Hawker Siddeley Aviation. From 1977 to 1993 he worked

Scott Crossfield (NASA)

as an aviation technical committee member for the US House of Representatives, which asked him to assist in investigating the loss of the space shuttle Challenger. He was killed when his Cessna 210A crashed in a thunderstorm in Georgia on April 19, 2006. Crossfield was awarded the Clifford Harmon Trophy for 1960 and the Collier Trophy for 1961 among many other honors.

Launched at 40,000ft and Mach 0.75, it was meant to climb at 38 degrees, with its engine burning for 75 seconds to propel it ballistically to 700,000ft (over 132 miles) before it dived at speeds approaching Mach 9. Any minor corrections to the aircraft's direction would have to be made during those hectic moments of acceleration, as there would be little opportunity to rectify errors and ensure a safe, spiraling, gliding return to the Muroc runway thereafter. The calculations involved in these flight plans and in the operation of the X-15 provided valuable experience for the space programs and space shuttle flights that would follow.

Faced with such extreme heat, the pilot's safety was a major concern and the cockpit was given a 2in.-thick layer of insulation while the windshield consisted of two layers of 0.5in.-thick quartz with a 0.25in. gap between them. There was also doubt about exposing a pilot to the cosmic radiation that he would experience in the ionosphere, but no additional protection was provided in view of the short exposure times planned for each flight. Designers even

D-558-2 NASA 144 is seen here on 26 October 1954 with NASA's test B-47. Sharing of data on swept-wing research was of great benefit to the US aviation industry and it enabled Boeing to build its first swept-wing bomber, the B-47, and many subsequent generations of swept-wing bomber and transport aircraft. (NASA via Terry Panopalis)

considered and rejected the potential risks from meteorite strikes, an entirely new problem for pilots of any manned aircraft. Although the ejectable nose section gave better protection to the pilot than previous versions, it was acknowledged that it could only be used over a part of the flight profile. At the extremes of speed and altitude, its potential effectiveness was unpredictable.

The US Navy strongly supported the Douglas design, but the Project 1226 hypersonic aircraft competition was won by a somewhat reluctant North American Aviation, which had no previous rocket plane experience. For its winning X-15, the company specified a massive XLR-99 rocket engine (an XLR-30 version nine times more powerful than the D-558-2's engine), Inconel-X chrome-nickel structure and a similar straight wing to the Model 684. Inconel-X material was preferred by the selection panel and it became a main factor in the success of the X-15 design.

Like Bell, Douglas had accumulated the bulk of US experience in developing rocket-powered flight, so it was unsurprising that many of Douglas's ideas, including those on high-altitude hydrogen-peroxide control systems, were used in the final X-15's conception. The USN was eventually given only a token role in the ground-breaking X-15 project, despite its Douglas design coming a close second in the Project 1226 competition when the result was declared on September 30, 1955. It was also ironic that, as the space program progressed from high-altitude flights in X-planes and the D-558 to landings on the Moon, seven of the 12 astronauts who walked on the Moon came from naval backgrounds, including six of the seven commanders of Apollo spacecraft that made lunar landings.

The X-planes were necessary to explore areas of high-speed flight that could not be simulated in contemporary wind tunnels. However, the existing tunnels were in turn essential for research into characteristics such as an aircraft's likely behaviour in a spin. This spin-test model was one of many used to study airflow patterns in NACA's Langley facility in May 1947. (NASA via Terry Panopalis)

For the same Project 1226 competition, Bell's D-171 proposal was based on its swept-wing X-2 and used three 14,500lb-thrust XLR-81 rocket motors. They burned extremely hazardous fuming nitric oxide as the oxidizer for their fuel system so that there would be no need for frequent topping up – a complex process during the airborne carrier phase that was needed by liquid oxygen-powered rocket planes. A massive Convair B-36 would have been its carrier aircraft. Bell opted for an innovative double wall method with two layers of Inconel-X skin separated by an air gap to provide insulation.

The similarly unsuccessful Republic AP-76 proposal was a heavier design that also depended on a B-36 carrier and XLR-81 engines, four in this case. It drew upon elements of the company's unsuccessful XF-103 Mach 3.7 interceptor proposal, having no conventional cockpit canopy so that the pilot's vision was limited to tiny side windows and a periscope for forward vision at low speeds. It came last in the competition.

North American Aviation rolled out its first of three X-15s in October 1958, and at the ceremony Walt Williams spoke of the debt owed by its designers to those who worked on the X-1 and D-558. Many of those present were also aware of the importance of Scott Crossfield's input to its design. As the first X-15 pilot, Crossfield became very involved in the engineering, data processing, and wind-tunnel testing of the aircraft. He had made a similar contribution to the D-558-2 when he became involved with it in 1951, particularly in the months leading up to the first Mach 2 flight, and one of his colleagues stated that he was effectively project engineer on the Skyrocket at that time.

CHAPTER SIX

AFTERTHOUGHTS

Showing a strong D-558-2 influence, Grumman's F11F-1 was the US Navy's first supersonic aircraft and the first to be designed around the "area rule" concept which considerably reduced transonic drag. Grumman chose a low-mounted horizontal stabilizer position. The F11F first flew in July 1954 and entered limited production, seeing service as a lightweight day fighter equipping six squadrons. This example, with folded wingtips, is seen on board USS *Forrestal* (CVA-59) in April 1956. (US Navy)

The D-558 aircraft were truly pioneering research tools that contributed greatly to the jet fighters and commercial aircraft of the 1950s. In direct practical terms, the experiments with adjustable tailplanes conducted by the Skystreak and Bell X-1 allowed later models of the F-86 Sabre to be fitted with variable-incidence stabilizers, which helped to prevent pitch-up and improved the fighter's maneuverability in combat during the Korean War. Understanding the correct positioning of horizontal stabilizers to alleviate pitch-up problems influenced the design of the X-15 with its low-mounted stabilizer, and enabled it to contend with the severe angle-of-attack situations it would face in re-entering denser atmosphere after flights above 250,000ft.

Research into buffeting, longitudinal stability, and aileron effectiveness were vital in formulating successful designs for the Century Series of supersonic fighters. Although the Skyrocket reached Mach 2, its pilot, Scott Crossfield felt that its main contributions in solving pitch-up problems were made "over a speed range up to about Mach 1.5. We never could do much when we got above those speeds because the speed wouldn't stay up there long enough to maneuver and do what we wanted." Of his own record Mach 2 flight he said in 1998, "It means nothing technically. It meant nothing from a research standpoint. It only meant that we got in the ball game and we got a score on the board. And we beat Yeager there that time."

Although wind-tunnel design had advanced considerably during the nine years of D-558 flying, the six aircraft and their USAF stablemates had shown the necessity of flying test-beds to prove

the accelerating technology of the time. The speed and altitude records broken by the D-558s were impressive, but their main contribution lay in the methodical accumulation of solid data on the fundamentals of high-speed flight, which enabled the design of a new generation of jet aircraft. NACA's instruments measured pressure distribution and wing loading over five points on the Skyrocket's wingspan, with others in the tail so that the relative efficiency of its airfoil sections could be studied in the transonic speed range. Engineers were provided with reliable information on loads measured at various points on the airframe so that they could evaluate the function of each part of an airframe more effectively during level and maneuvering flight. Improvements in understanding factors that induced instability in the transonic range were one of the most valuable outcomes, although the inherent problems would persist until complex auto-stabilization systems became available for production aircraft.

In terms of comprehensive research data acquisition, the D-558 series yielded great benefits. The pilots who often risked their lives to investigate the dangerous pitch-up tendencies at the supersonic boundaries provided particularly valuable knowledge. The D-558s and the parallel Bell X-1 series proved beyond question the value of purpose-built experimental aircraft coupled with well-organized scientific back-up to interpret and use the research data that they generated. In the case of the D-558-2 this was also completed without loss of any pilots or aircraft, despite a long and intensive trials program – an unusual achievement in test flying at that time. From 1948 to 1954, when the D-558s were making their fullest contribution to transonic research, NACA's High-Speed Flight Research station at Muroc grew from a very basic functional facility to a complex and well-funded organization with six times as many staff and ambitious plans for hypersonic vehicles and space travel.

The second Skyrocket, with one rocket chamber lighting up, launches from its P2B-1S transporter. The rocket motor could often be lit as the Skyrocket left the P2B-1S's bomb-bay shackles, and it would start to power the rocket-plane's flight within a hundred feet from the drop. NACA found that the D-558-2's drag coefficient was "somewhat less than the X-1 airplane," making it fly better immediately after launch. The aircraft's surfaces were frequently repainted and polished. After the first flight at Mach 1.36, the whole aircraft was repainted with 15 coats of white lacquer as much of the original finish had been worn off. (NASA)

FURTHER READING

BOOKS

Baker, David, *Rocket Planes* (Aviation Classics, Horncastle, 2017)

Bridgeman, William, and Hazard, Jacqueline, *The Lonely Sky* (iUniverse, Inc., New York, 2009)

Crossfield, Scott A., with Blair, Clay, Jr., *Always another Dawn* (Hodder and Stoughton, London, 1961)

Davies, Peter E., *Bell X-1*, X-Planes No. 1 (Osprey Publishing, Oxford, 2016)

Davies, Peter E., *North American X-15*, X-Planes No. 3 (Osprey Publishing, Oxford, 2017)

Davies, Peter E., *Bell X-2,* X-Planes No. 6 (Osprey Publishing, Oxford, 2017)

Everest, Frank, with Guenther, John, *The Fastest Man Alive* (Cassell, London, 1958)

Forsyth, Robert, *Bachem Ba 349 Natter*, X-Planes No. 8 (Osprey, Oxford, 2018)

Gorn, Michael H., *Expanding the Envelope* (The University Press of Kentucky, Lexington, 2001)

Hallion, Richard P., *Supersonic Flight* (The Macmillan Company, New York, 1972)

Libis, Scott, *Douglas D-558-1 Skystreak*, Ginter Naval Fighters No. 56 (Steve Ginter, Simi Valley, California, 2001)

Libis, Scott, *Douglas D-558-2 Skyrocket,* Ginter Naval Fighters No. 57 (Steve Ginter, Simi Valley California, 2002)

Yeager, General Chuck, and Janos, Leo, *Yeager* (Century Hutchinson Limited, 1986)

DOCUMENTS

Hallion, Richard, "On the Frontier – Flight Research at Dryden, 1946–81," NASA Publications, SP-4303, 1984

Love, James E., and Stillwell, Wendell H., Technical Note D-185, NASA Dryden, 1961

NACA Research Memorandum, "Wind-tunnel Investigation of the Stability of Jettisoned Nose sections of the D-558 Airplane," Stanley Scher, Langley Memorial Aeronautical Laboratory, Washington, DC, January 1948

Hypersonics Before the Shuttle (Dennis R. Jenkins), NASA Publications SP-2000-4518, June 2000

Towards Mach 2: The Douglas D-558 Program, NASA SP-4222, Washington, DC, 1999

INDEX

References to images are in **bold**.

accidents 7, 28–29, 30, 39
air intake 9, 12, 15, 34, 42
aircraft, British 7
    Boeing B-29; 8
    S.6B seaplane 5
aircraft, German:
    Heinkel He 178; 6
    Me 163 Komet 9
aircraft, US:
    Bell Airacomet 6
    Bell X-2; 49
    Bell XS-1; **8**, 9, 13–14, 17, 20, 24
    Douglas D-558-1 BuNo **18**
    Douglas MX-656; 68–69
    F-86 Sabre 77
    Model 671; 68
    Model 684 "Skyflash" 71–75
    P2B-1S (*FERTILE MYRTLE*) **49**, 50,
        51, 52, 64–65
    X-3 Stiletto 58, 69–71
    X-15; 75, 76, 77
    see also Douglas D-558-1 Skystreak;
        Douglas D-558-2 Skyrocket
altitude 55, 56–58, 61, 62–63, 72–73, 78
Ankenbruck, Herman O. 59–60, 65
Armstrong, Neil 65, 66–67
Arnold, Gen Henry H. "Hap" 4, 6

Bell, Larry 24
Bellman, Donald 32
bombs 5, 66, 67
Boyd, Col Albert 14
braking parachute 16
Bridgeman, William Barton "Bill" 41, 42,
    43, **66**, 69–70
    and air-launching 52, 54–58, 59
Briggs, Dr Lyman 4
Busemann, Adolph 5, **7**
Butchart, Stan 15, 30, 64–67

Caldwell, Frank 4
Caldwell, Cdr Turner F., Jr **4**, **9**, 21, 25
canopy 15, 21, 29
Cardenas, Maj Roberto 56
Carder, Al 55
Carl, Lt Col Marion **4**, **19**, 21, 25, 61–64
Carroll, Brig Gen Franklin 8
Champine, Bob 17, 24, 30–31, 43–44,
    45
cockpit 15–16, 41, 74
    and Skyrocket **53**, **55**
compressibility 4–5, 6, 14
Conlon, Cdr Emerson 14
cooled air 17–18
"Crimson Test Tube" *see* D-558-1
    Skystreak
Crossfield, Albert Scott 17, 30, 31, 32,
    39, **46**, **74**
    and air-launching 59, 60, 64, 65
    and escape system 41
    and hypersonics 76
    and pitch up 77
    and Stiletto 69, **70**

De Havilland, Geoffrey 14
Derry, John 14
Donaldson, Grp Capt Edward "Teddy" 14

Douglas D-558-1 Skystreak **4**, 7–8, 9,
    12–14, **26–27**, 77–78
    and mock-up 15–17
    and NACA 24–25, 28–31
    and speed records 21, **22–23**, 24
    and stability 31–32
    and testing 18–21
Douglas D-558-2 Skyrocket **10–11**, 14,
    33, 34–35, **36–37**, 38–39, 77–78
    and air-launching 49–52, 54–67
    and building 39–48
    and cockpit **53**
Dryden, Dr Hugh 4, 49, 64
Dutch roll 43, 54, 57

ejection seat 14, 70
elevators 31–32
endplates **18**, **20**, 25
escape 14, 17, 40–41; *see also* ejection seat
Everest, Lt Col Frank "Pete" 65

Fales, Elisha 4
fuel 19, 38–39, 40, **68**
full-pressure suit 57, 61, 62–63

Germany 33–34; *see also* aircraft, German
Griffith, John 4, 30, 34, **38**, 44–45

Heinemann, Ed 12, 14, 15–16, 67, 72
Hilton, W. F. 5
Hoover, Herb 30–31
Hyatt, 1Lt Abraham 8, 14
hydrogen peroxide 38–39, 51, 72–73
hypersonics 71–76

instrument panel 15–16, **18**, 28, 51
ionosphere 72, 74

Jansen, George 25, 54, 55–56
jet-assisted take-off (JATO) 42–43, 54
jet engine 41–43
Jones, Robert T. 5
Jones, Walter 59, 60

Kármán, Dr Theodore von 4–5, **6**, 8
Klein, Col 20–21
Korean War (1950–53) 77
Kotcher, Ezra 5–6, 7, 8, 9, 33

landing gear 18, 19, 28, 40
Lewis, Dr George 5, 6
Lilly, Howard "Tick" 7, 28–29, 30

McKay, John 30, 32, 58, 65, 67
Martin, John F. "Johnny" 25, 42–43
May, Eugene F. "Gene" **15**, **16**, 19–20,
    21, 25, **30**
    and records 24
    and Skyrocket **35**, 43, 46–47, 48, 54
Moon landings 75

NACA (National Advisory Committee for
    Aeronautics) 4, 7–9, 12–13, 78
    and air-launching 49, 52 57, 57,
        58–60, 64–65
    and deal 24–25, 28–31
    and mock-up 16, 17
    and Skyrocket 34, 47–48
    and tests 31–32

nitrogen flushing system 51–52
nose section 16–17

oscillograph 18, 41

Peyton, Larry 25
pitch up 44–45, 60, 77

radiation 74
Ridley, Lt Col Jack 48
rocket motors 7, 34–35, 38–39, 46–48
Rocket Servicing Trailer 39
Root, L. Eugene 14

shock waves 31, 61
Smith, A. M. O. 14
Smith, Robert J. 12
Smith, Stan 34
Soucek, Rear Adm Apollo 64
Soulé, Hartley 49
speed 4–7, 8, 12, 13–14, 20
    and air-launching 55–57, 61, 63–64,
        65
    and records 21, **22–23**, 78
    and Stiletto 70
stability 31–32, 60, 78
Stack, John 5, 6–7, 15, 33
supersonic speed 4, 5–7, 8, 13, 33
    and Skyrocket 47, 48
Synasol 38–39

tailpipe 13, **20**, 40
take-off 42–43, 47–48
tanks 38
Tipton, Joseph 65
transonic speed 7, 8
Trapnell, Capt Frederick 25
Truszynski, Gerald 17
turbojet propulsion 7, 8, 38

USAAF (US Army Air Force) 7, 8, 48
USN (US Navy) 6, 7, 8, 9, 12
    and bombers 50
    and hypersonics 75
    and Skyrocket 46, 48
    and wings 34

Van Every, Kermit 40
Vensel, Joseph R. **46**
Volta Congress on High Speeds in
    Aviation (1935) 5

Walker, Joe 30, 65, 70
weight 40
Welch, George 33
wheels 18, 40
Williams, Walter C. "Walt" 25, **46**, 69,
    70, 76
    and air-launching 61, 64
wind tunnels 4, 5, 6–7, 9, 12, 77–78
wings 12, 13, 40, 60–61
    and sweep 5, 7, 33–34
Wolf, Robert A. 7
Woods, Robert J. 9

Yeager, Capt Chuck 15, 24, 33, 48
    and air-launching 56, 65

Zornig, Lt Col H. 5